Signals and Communication Technology

More information about this series at http://www.springer.com/series/4748

Asoke Kumar Datta

Acoustics of Bangla Speech Sounds

Springer

Asoke Kumar Datta
Society for Natural Language Technology
Research (SNLTR)
Kolkata, West Bengal
India

ISSN 1860-4862 ISSN 1860-4870 (electronic)
Signals and Communication Technology
ISBN 978-981-10-4261-4 ISBN 978-981-10-4262-1 (eBook)
DOI 10.1007/978-981-10-4262-1

Library of Congress Control Number: 2017937264

Printed on acid-free paper

This Springer imprint is published by Springer Nature
The registered company is Springer Nature Singapore Pte Ltd.
The registered company address is: 152 Beach Road, #21-01/04 Gateway East, Singapore 189721, Singapore

To my dear wife, late Suvra Dutta

Prologue

Categorization of basic units of speech sounds in a language may be done using two different approaches. One is the traditional way of subjective judgment of linguists/phoneticians. The other is the objective one which is based on the acoustic analysis of the corresponding speech signals collected from many native speakers. The two are complementary to each other. The former may have an element of inconsistency due to personality bias and one is not sure that such characterization is really *consonant* with the speech of the native speakers. Though objective analysis may be free from these shortcomings it has to lean heavily on the subjective judgment regarding categorization simply because speech is an intelligent subjective phenomenon. In between pure subjectivity and the pure objectivity *of pointer reading* in hard science there is a place for public subjectivity which grows out of common perception of many people and when they could be correlated with some objective measurements one could have a trustworthiness associated with scientific propositions. We may call it objective subjectivity. It blends seamlessly these two representations of reality. This underlies the subject matter in this book.

The burgeoning demand of technology development in speech needs objective parameters to be drawn from the speech signal for each of these basic units as well as their mutual interactions. These signals should ideally be drawn from the speech of the native speakers. This need for technology is not extravagant in Indian socio-economic context rather it is of prime importance. This is intrinsically related to the need to empower the people in the socio-economically disadvantaged populace in India. Knowledge is power. Thanks to the tremendous development in the information technology in the last few decades, the knowledge resources at the global level in respect of all conceivable disciplines relevant to human development, particularly in the field of common productivities like farming, fishery, etc. is available in the e-domain. Unfortunately most of these disadvantaged people are functionally illiterate and therefore cannot directly access these sources of knowledge. One may note that speech mode is the most common and handy medium for knowledge transfer particularly amongst this segment. The oral transmission of knowledge is still the force majeure. The Spoken Language Technology (SLT) as the most favored means for knowledge transfer to achieve inclusive development in the developing and underdeveloped countries should be seen in this perspective.

The traditional studies on acoustic phonetics of Bangla speech sounds, the standard colloquial dialect used in West Bengal, India, have a long history and are based generally on the perception of erudite scholars, which as mentioned earlier are subjective in nature. Apart from the historical perspective, these have significant relevance as a knowledge source still today. However, as a spoken language is highly dynamic and constantly evolving there is a need to update the knowledge from time to time. Apart from this, nowadays, in addition to the scholarly pursuit of knowledge there is a technological need to fathom the objectivity of speech sounds. This need arises out of the aforesaid urge of empowering functionally illiterate people of the country to the vast knowledge bases, general as well as specific profession oriented, existing in the electronic digital domain. The development of speech technology for this purpose needs objective categorization and parametric representation of these sounds. While some work has been done in this direction in the last three decades these need to be revisited in the light of developing technology for objective assessment and to be consolidated so that a comprehensive state-of-the-art representation is available at a single source.

The paradigm used here is to analyze real speech of selected native speakers of the dialect Bangla. This is done, apart from the scholarly need, to ensure that a proper acoustical data base is available for the development of speech technologies to be used by a commoner for the aforesaid empowerment. Statistical tools are available to get proper representative values for each of the acoustic parameters from a collection of variety of utterances of a particular speech unit, varied due to variation of speakers and of contexts, as well as a measure of the variability. These are imperative for the development of speech technologies. It may not be out of place to mention just the two important speech technologies for the empowerment of the commoners. These are text to speech synthesis (TTS) and automatic speech recognition (ASR). TTS is needed for machine-to-man communication in speech mode. This makes digital knowledge base directly available to a human recipient in normal speech mode. Indigenous TTS engine has been developed as early as 90s. This is now available as a product form. ASR is primarily, in the aforesaid context, required for the machine to understand the query presented in oral mode. This is a more difficult task even in limited domain. As regards ASR, though quite sophisticated systems are available for many foreign languages, the indigenous development has not reached the appropriate level for general public use in any of the Indian dialects. However, such indigenous attempts are currently being tested in very restricted domain in some of the Indian dialects.

Several instrumental setups are necessary for the extraction of the objective parameters directly from the speech sounds. Some of the instruments used in this field are electronic palatograph (EPG), electroglottograph (EGG), spirometers, cine-fluorography, ultra-sonography, sound spectrographs, endoscopy and video graphs. It is also possible to use sound spectrography to estimate some objective properties in the absence of a particular important instrumental setup. One needs to understand that these instrumentations are necessary only to objectify the categorization of the units of speech. Obviously these cannot be used in the aforesaid technologies at the practical level. The information gleaned from these instruments is used in speech signal processing. One must emphasize here that only the sound waves are available for the TTS or ASR engines at the ground level. In this context the speech signal processing (SSP) becomes the most important tool in the armory of speech technology mentioned above, and the acoustic data base is the most important component of it. With this objective in view, the dialect of Standard Colloquial Bengali (Bangla) is chosen as the base spoken language for acoustic parameterization in this book. It may not possibly be out of place to mention here that the availability of comprehensive as well as extensive field data base of acoustic information in any Indian dialect at a single place is not available up-to-date.

Asoke Kumar Datta

Acknowledgements

It is difficult to recollect all who have contributed in the research for acoustic and other information spanning over three decades, particularly when age begins to nibble at the memory. One could only wish to be forgiven for any unintended failure to remember all and express gratitude to them. I must begin with mentioning the name of Prof. Dwijesh Dutta Majumdar, a renowned scientist in the field of computer and electronic communication science and the-then head of the Department of Electronics and Communication Sciences, Indian Statistical Institute, where the research began and that of Sri Amiya Baran Saha's, the-then Executive Director, CDAC, Kolkata, where the investigations finally bloomed. During my investigation in CDAC many faculty members, research scholars, students and project assistants contributed their time and knowledge. Of them I remember late Bijon Mukherjee, Sarvasri Nihar Ranjan Ganguly, Anuradha Roy, Somnath Das, Tarun Dan, Somen Chowdhury, Shyamal Das Mondal, Arup Saha, Tulika Saha. I acknowledge their contribution with a note of thanks. I also thank all scholars, project assistants, informants and other incumbents, who have helped at different times in the studies over the spanning decades with an expression of regret that I do not remember all the names.

Contents

Chapter 1
Vowels, Glides and Diphthongs

1.1 Introduction

Bengali is the official state language of the Eastern Indian state of West Bengal and the national language of Bangladesh. It is a part of the *Indo-Aryan* (IA) branch of the Indo-European family of languages (Fig. 1.1). Mutation of its spoken form over the 1000 years after breaking away from Bihari has given rise to at least seven distinctly different dialects in West Bengal and at least four dialects in Bangladesh (Fig. 1.2). It is also spoken in the states neighboring West Bengal, that is, Assam, Bihar, Chattisgarh, Tripura, and Orissa as well by the Bengali Diaspora around the world. It is true that the concept of dialects, particularly the division of the speech of an ethnocultural group, into different speech classes or dialects is always controversial. However societies and nations have always been doing that for the convenience of societal governance which is by nature locally stationary. By locally stationary, what is meant is that for a period of time people agree with these divisions. This is one particular point one has to keep in view when trying to formalize a speech system. One can have relatively easier task with written languages because the units are more crisp and robust and are stationary. For speech the units are fleeting. Not only that the units are liable to have instantaneous mutations due to anticipatory and/or correlatory influences, it has to be borne in mind that, at least for any dialect under Bengali, not much objective knowledge related to the speech of native speakers has seen the light of the day.

For obvious reasons the dialect spoken in the *south east* part of Bengal has been chosen as the standard colloquial form of Bengali (SCB) and is also referred to as Bangla. This is the form that is being used generally, in radio, TV broadcasting as well as formal official communication.

Since the pioneering work of Suniti Kumar Chatterji in early twentieth century, a large number of eminent linguists of West Bengal and Bangladesh have contributed to the development of Bengali phonetics based on subjective perception (Hai 1989; Chatterji 1926). According to these literatures the non-consonantal phoneme sets

A.K. Datta, *Acoustics of Bangla Speech Sounds*, Signals and Communication Technology, DOI 10.1007/978-981-10-4262-1_1

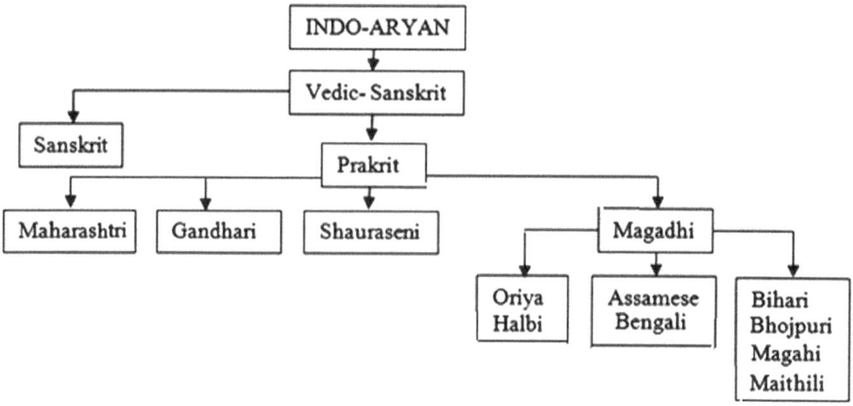

Fig. 1.1 Position of Bengali in the Indo-Aryan family

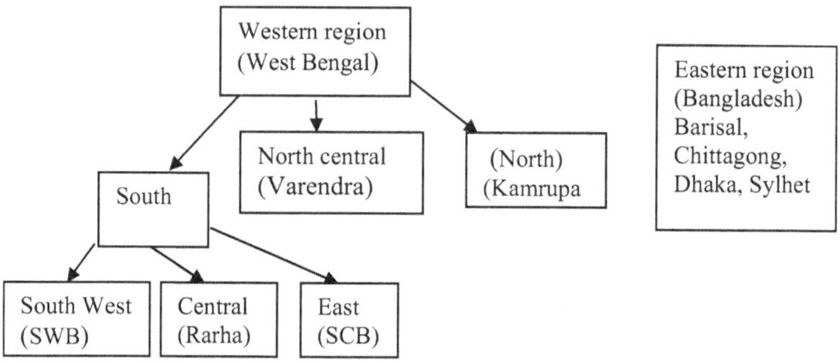

Fig. 1.2 Basic dialect profile of spoken Bengali (see Chatterji 1926; Varshney 1995)

used in Bengali speech consists of 14 vowels (oral and nasal), around 20 diphthongs (number vary according to authors) and *3* semi-vowels.

All the categorization and the description of the aforesaid phonemes related to Bengali speech are based primarily on the perception by the erudite scholar of a particular item spoken by a native speaker. Two lacunas come to the mind. The first is how one chooses a representative utterance of particular phoneme from a large number of such utterances by speakers of different sexes or ages. The procedure has never been widely reported. The other one is whether the perception could be biased by some favorite model inside the mind of the erudite author. To what extent these reflect the actual practice of the people native to the dialect is a matter of individual opinion. However in the modern era the advent of technology provides facilities for objective measurement and a global shift in the paradigm of characterizing phonemes has taken place. The emphasis is now on analyzing objectively real speech of a large number of native speakers of a dialect to arrive at the appropriate representative categories. Some are based on analysis of acoustic parameters, some use imaging technology which includes cine-radiogragphy, ultra-sonography, electronic palatography and the like. The idea is to define individual speech events in terms of quantifiable physical measurements. Highly developed statistical techniques are used to find representative of the required speech event which fits best the observations from native utterances. As there exist individual variations, it is also necessary to have their measures included in the data base. Fortunately statistics come handy here too.

It is even possible to use clustering techniques to find the number of separate elements, say for example the system of vowels in an unknown tribal language. This in no way interferes with researchers who prefer to use skillfully honed ear and cognitive abilities for analyzing exceptional situations or where semantics is involved. The objectivity is necessary because of the vast potential of technology development for empowering people to access knowledge fast. In India this approach was gaining ground since early 70s (Djordje and Das 1972). Some work in this vein has also been reported on the dialect SCB on a small data base (Ganguly et al. 1988, 1999; Datta et al. 1989). The instrumental analysis may help sometimes in resolving confusion like whether some Bangla vowels like /i/, /u/ really have long and short counterparts like those in Telugu (Datta et al. 1978), or resolving the controversy over the number of diphthongs in Bangla and the like.

However one must note here the limits of this objective analysis. One is the limitation of the instruments. For example the EPG analysis cannot give concrete evidence of velar closure. Similarly the differentiation between voiced and unvoiced manners of articulation cannot be fully determined from the acoustic evidence. The other is the cognitive one influenced by context as well as semantics. For example the acoustic parameters of the nucleus vowel in the word /sɐt/ (meaning 'seven') is that of the vowel /æ/ but cognitively it is /ɐ/.

The book uses the objective paradigm of investigation with a reasonably large data base and larger number of informants along with the present state of art technology for the objective parametric evaluation. Along with it the conformity of these parametric evaluations with the traditional or otherwise subjective categorization is maintained throughout.

1.2 Vowel, Semi-Vowels, Diphthongs

Vowels, semi-vowels and diphthongs consist one group of fundamental units of a spoken language characterized by the periodic structure produced by oscillations of the vocal cords in the glottis where the air path is unobstructed. There are other fundamental units which are also characterized by periodic structure and vocal cord oscillations like murmurs, e.g. /ŋ/, /ɲ/, /n/, /m/ and voice bars in voiced consonants and voiced fricatives like /z/ but these are characterized by full or partial obstruction of the air passage. However as these are grouped as consonants these will be considered in a later chapter. The traditional subjective categorization of each unit in the former group can be distinguished by their characteristic timbral quality and the dynamic behavior of this quality. This cognitive unit, timbre, has a well defined physical dimension called complexity. The complexity again is measured by the spectral structure of the sound. Generally for speech sounds, spectral structure refers to resonances and anti-resonances which respectively refer to hill and valley like structures in spectra (see Fig. 1.7, Sect. 1.3.1). These resonances are characteristic of the shape and sizes of resonating cavities contained in oropharynx and in some cases also nasopharynx. The first two resonances called formants, F_1 and F_2 (Fant 1970) are often used for the characterization of vowels. F_1 and F_2, roughly correlate with the tongue height and tongue position respectively. Using F1 and F2 it is possible to properly place them in a vowel diagram. However we shall see in Sect. 1.3.2 that it is also possible to estimate the height and backness of the tongue hump using F_0 along with F_1 and F_2 and place the vowels more appropriately in the vowel diagram. The aforesaid height and backness are again other estimates done cognitively by the phoneticians. There has been no purely hard technology solution for regular application simply because the only way to measure these is to use X-rays which is hazardous and therefore cannot be resorted to. The other expensive alternative is computer tomography.

While the vowels are characterized by a steady timbre of cognitively relevant duration the semi-vowels are sounds perceptually distinctive in the sense that they appear to have a cognitive gliding movement of the sound and so are also referred to as glides. Being a syllable nucleus it also has a target position. Jones (1962) described semi-vowels as a voiced gliding sound in which speech organs start by producing a vowel of comparatively small prominence and immediately change into a more prominent vowel. It is suggested that the duration of either the on-glide or the off-glide must be comparable to that of the target (Lehiste and Peterson 1961). This movement can be seen from the movement of the formants in a spectrograph and sometimes in the pitch contour. Ladefoged described semi-vowels as an approximant consisting of a non-syllabic vowel occurring at the beginning or the end of a syllable (Ladefoged 1967). While these descriptions appear to be more perception based they do have acoustic verifiability and corresponding objectivity. It is a matter of choice for the investigator what paradigm to choose.

Diphthongs are syllable nuclei, which are perceived as two vowel targets. One of them appears significantly stronger than the other. Acoustically they can be seen as a spectral movement ending on one side into a cognizable duration of steady state representing a vowel and on the other side just a target for a different vowel having very little or no duration (Hossain et al. 2005).

Speech is man-to-man communication, and of course it is purely subjective. At the same it was found that it is possible to bring in strong objectivity (Fant et al. 1972). However bringing objectivity in a primarily subjective affair is no mean task. The problem is also psychoacoustic in nature. Let us take the example of a glide. For a particular piece of glide everybody may not actually hear it. One may decide to take the mean value of duration for which most of the listeners hear the glide. One could design an elaborate listening experiment for the purpose. One could also collect a large number of spoken word containing a glide and glean the knowledge from it. We have seen above that the glide is defined only as formant movement in a word. The problem with real speech data is that human speech is not concatenation of separately spoken words. There are events between two words primarily caused by co-articulation following the need of continuity. We also shall see that there are perceived glides where formants play almost no role, pitch contour takes over.

1.2.1 Data Base

The collection of speech data from native speakers is a complex and arduous task on its own. Sometimes specific efforts are necessary depending upon the objective. Fortunately such an exercise has been done recently by CDAC, Kolkata and DIT, Govt. of India has made it free for use in research and technology development. The samples needed for analysis of vowels, semi-vowels and diphthongs for the present work have been taken primarily from these sources. The speech samples for the present study consist of 595 Phonetically Balanced Word (PBW) set. PBW is a special set of words designed such that it contains all possible CV and VC combinations available medially in dictionary words. It also attempts to ascertain that the numbers of each syllable in the collection are reasonably similar. This requires an elaborate trial and error paradigm in conjunction with the use of an excel sheet and the Samsad Bangla Dictionary (2004) (used for the purpose). An examination of this set revealed that many semi-vowels and diphthongs are either not available or the numbers are not adequate for a statistical representation. It was therefore decided to augment the list. In this process 195 additional words had to be inducted from Bangla Samsad Dictionary (2004) to include such semi-vowels and diphthongs those do not find place in the PBW list. All the words thus selected are embedded in a neutral carrier sentence [ami ebar '….' boltʃʰi] for recording. Thus the prosodic environments of the words are expected to be the same.

The informants were all native speakers of Bangla and their ages were between 20 and 50 years (Table 1.1). The numbers of informants for vowel recording are 4 male and 4 female whereas those for diphthongs and semi-vowels are 2 for each sex. The process of selection of informants again is an important task. The number

Table 1.1 Meta data for informants

Speaker	Age (years)	Male/Female
1	23	Male
2	25	Male
3	36	Male
4	40	Male
5	24	Female
6	25	Female
7	36	Female
8	42	Female

of informants is usually constrained by the availability of resources. In such cases extra attention has to be given on the process of selection. A large number of native educated speakers from the selected dialect were invited to participate in an audition procedure. A list of words, both sense and nonsense along with a selection of some reading material was prepared for the purpose. The material has to be subdivided into sections for reading to give adequate rest to the speakers in between. These readings were then examined by a committee comprising of experts from the fields of linguistics, dramatists and speech technologists for final selection of informants.

The recording of the sentences corresponding to the additional words was done using the Cool Edit Pro software in speech studio environment (reverberation time 0.4 s) in normal text reading mode. The recording format is 22050/16/mono wav. A session was of 5 min duration followed by a 5 min break. The environment and format is same as that of speech data recorded in CDAC corpus. After each recording, the moderator checked for any error during the recording, and if so, the erroneous sentence was recorded again. During the recording a video of the informant's frontal face was also captured.

A total of 800 segments for each Bangla vowel and 479 segments for diphthongs and semi-vowels of the said native informants was taken for the study on semi-vowels and diphthongs.

1.3 Vowels

The study of phonetic and articulatory characteristics of Bangla vowels by some eminent linguists of both West Bengal and Bangladesh (Hai 1989; Chatterji 1926) reveals that there are altogether 14 vowels (both oral and nasal) in Bangla namely /ɔ/, /a/, /i/, /u/, /æ/, /e/, /o/, /ɔ̃/, /ã/, /ĩ/, /ũ/, /æ̃/, /ẽ/, /õ/. It is also noted there the existence of a separate short /ʌ/ sound found only in loan words (Chatterji 1926). These categories relate to manner (oral/nasal) and place of articulation. The later is associated with the place and degree of the constriction which divides the oral cavity into two. When the tongue is the articulator the horizontal position of the tongue hump as well as its height defines the class of the vowel. The degree of rounding is important for the articulator lip. Table 1.2 presents the oral Bangla vowels according to the traditional literatures.

The graphemes in rows 3, 4, 10 and 11 indicates shorter and longer counterpart of the same phoneme only in textual representation. However no consistent differentiation in utterances in Bangla has ever been reported.

Traditionally vowels are characterized by the height and position of the tongue. Figure 1.3 shows the position of the tongue for 8 cardinal vowels given by Jones (1962). This is the result of only direct objective determination of tongue position. This led to the primary cardinal vowel quadrilateral that is still used albeit with necessary evolution throughout the intervening years. The cardinal vowel system is intuitively related to articulation. The two-dimensional order of vowels according to tongue height and tongue position is immediately understandable and helpful, but it

Table 1.2 Traditional place and manner of articulation of oral Bangla vowel

S. No.	Grapheme	IPA symbol	Example Bangla words			Manner and place of articulation
			Initial	Medial	Final	
1	অ	**/ɔ/**	*/ɔsim/* 'Infinity'	*/kɔtʰɐ/* 'Talk'	NONE	Low mid back rounded (oral)
2	আ	**/a/**	*/ɐmi/* 'I'	*/kɐl/* 'Time'	*/kɐkɐ/* 'Uncle'	Low central unrounded (oral)
3	ই, ঈ	**/i/**	*/iliʃ/* 'Hilsa'	*/din/* 'Day'	*/nodi/* 'River'	High front unrounded (oral)
4	উ, ঊ	**/u/**	*/upɔre/* 'Up'	*/kukur/* 'Dog'	*/ʃɐdʰu/* 'Saint'	High back rounded (oral)
5	এ	**/e/**	*/ekʰɐne/* 'Here'	*/ʃebɐ/* 'Nursing'	*/kɔbe/* 'When'	Mid high front unrounded (oral)
6	ও	**/o/**	*/oʈʰɐ/* 'To get up'	*/lok/* 'Person'	*/ɐlo/* 'Light'	Mid high back rounded (oral)
7	অ্যা	**/æ/**	*/æk/* 'One'	*/ʈæg/* 'Abandonment'		Low mid front unrounded (oral)
8	অঁ	**/ɔ̃/**		*/gɔ̃d/* *'gum'*		Low mid back rounded (nasal)
9	আঁ	**/ɐ̃/**	*/ɐ̃kɐ/* *'to draw'*	*/kɐ̃tʰɐ/* *'hand stitched cover'*		Low central unrounded (nasal)
10	ইঁ, ঈঁ	**/ĩ/**	*/ĩdur/* *'rat'*	*/ʃĩdur/* *'vermillion'*		High front unrounded (nasal)
11	উঁ, ঊঁ	**/ũ/**	*/ũtʃu/* *'high'*	*/bũtʃi/* *'flat nosed girl'*		High back rounded (nasal)
12	এঁ	**/ẽ/**	*/ẽtʃuoɽ/* *'green grape-fruit'*	*/bẽʈe/* *'of short stature'*		Mid high front unrounded (nasal)
13	ওঁ	**/õ/**	*/õtʃʰɐ/* *'useless, rotten'*	*/põta/* *'to bury'*		Mid high back rounded (nasal)
14	অ্যাঁ	**/æ̃/**	*/æ̃ɽ/* 'a slang word'	*/pæ̃tʃɐ/* *'owl'*		Low mid front unrounded (nasal)

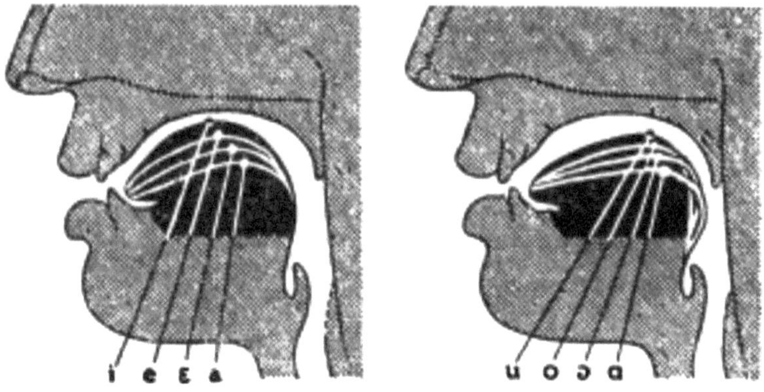

Fig. 1.3 Show the tongue position for eight vowels (Jones 1962: 32)

is only quasi-articulatory because the actual shape of the vocal tract which is three dimensional and irregular is difficult to be described by symbolic means alone. It may also be not necessary. One must note that the aforesaid classical work used elaborate and exhaustive X-ray analysis. This being hazardous to health this was rarely repeated later. The prevalent attempt is to describe the underlying symbols from articulation of vowels as exactly as possible. However in judging the articulation of vowels, one has to largely rely on personal assessments, which may oftentimes become vague. Furthermore when a linguist or a phonetician evaluates a phone he evaluates the one he considers correct. There is little guarantee that his utterance is really representative of the natives utterances. In the present day context of science and technology, issues involving language, such as phonemes along with their categorization, need to be seen in the context of language as a community object not as an exercise in ideational abstraction.

Apart from the aforesaid categories in Bangla vocalic sounds, namely oral and nasal, the existence of another manner has recently come into the ambit of investigation (Datta 2014). This is aspirated vowels. The existence of aspirated vowels has been noticed in many Indo-Aryan spoken languages. Unfortunately this has escaped the ambit of specific investigations for Bangla in the past.

Fant (1970) as early as late 60s and early 70s proposed modeling of the articulatory system using connected cylindrical resonators to represent oral cavities. His models, so far, has been found to be quite accurate, both in terms of spectral structure and in terms of cognition of synthesized vowels using Linear Electrical Analog (LEA). Since then spectral features have been considered to be the best approach for objective determination of the places of articulation of phonemes. One may refer to the early work of Delattre et al. (1952). In India it took about 15 years to adapt this development (Dutta Majumder and Datta 1966). However linguists and phoneticians, particularly in India, continued to depend on personal assessment for representing the vowels with symbols and their placement in vowel quadrilateral.

1.3.1 Method of Analysis

Before a discussion on the analysis of speech signal starts a brief introduction to the process of articulation of these sounds may be in order. Figure 1.4 presents a schematic representation of the articulatory processes involved. Air from the lungs sets in motion oscillations of vocal chords when the glottis is closed. These oscillations produce pressure pulses of air which in turn resonate the oropharynx. However if the velum happens to be open then the nasopharynx also resonates. Each and every cavity introduces its own resonance and anti resonance frequencies in a very complex manner. The tongue hump divides the oropharynx into front and back cavities. By changing the position of the tongue the brain controls the resonance structures in two basic ways. One is the relative sizes of these two cavities and the other is the coupling between them. Thus simply put the basic signal one needs to analyze for the present task is a repetitive signal whose complexity is controlled to put in the information along with its dynamics.

There is also a need for saying a few words about the repetitiveness of the signal. Traditionally the vocal cord oscillations were believed to be alike the vibration of a pair of rigid reeds. However it has been seen that the voiced speech is not completely periodic, it is quasi-periodic. This means that if we examine closely two consecutive periods they are not exactly alike. They differ randomly in time period (jitter), amplitude (shimmer) and complexity (complexity perturbation), though by tiny amount but good enough to provide a feeling of naturalness (Choudhury 2006). This quasi-periodicity is said to arise out of a model completely different (Teagre and Teagre 1990) from that of the vibration of rigid reeds. The mechanism is now considered to be more akin to flapping of flags than vibration of rigid bodies. It is now held that the muscle tissue of the vocal chords acts as rigid bodies (Fig. 1.5). It is the mucosal cover which really takes part in pressure pulse production. Air from the lungs forces open a free passage by shifting the mucosal layer apart letting a puff of air out, which reduces back pulmonic pressure and consequently the plasticity of the mucosal layer arranges itself back in a complete closure. This cycle repeats. The plastic nature of the mucosal surface introduces non-linear dynamics in the resultant air flow and a Kalman vortex stream is generated. This is responsible for the production of random perturbations referred to above.

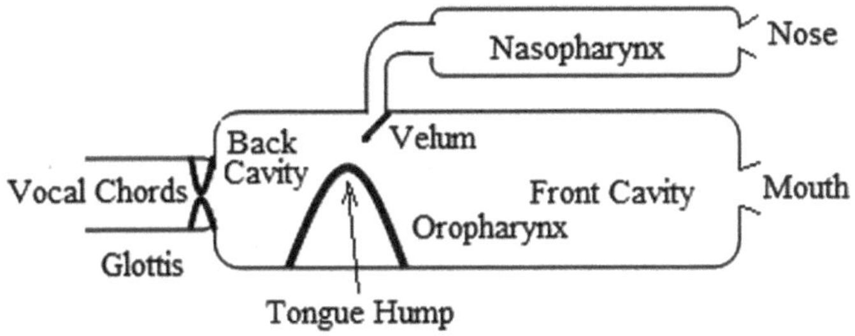

Fig. 1.4 A schematic diagram of voice production

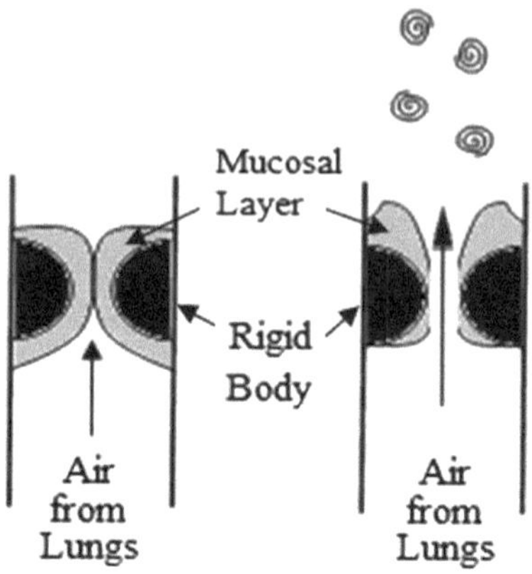

Fig. 1.5 Schematic of non-linear glottal oscillation

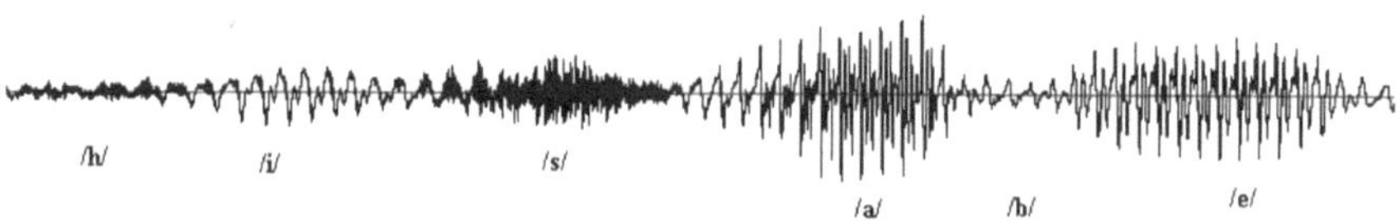

Fig. 1.6 An example of the signal wave form

Figure 1.6 presents the wave form representation of the signal for the word [*hisɐbe*]. It shows the wave forms of three vowels /i/, /ɐ/ and /e/ along with the signal of other sounds. The waveforms of the vowels are quasi-periodic and one way to characterize them is through their spectral structures while that for /s/ is quasi-random and the signal representing /h/ is a mixture of the two.

The spectral structure of /æ/ is shown in Fig. 1.7. The frequency and the amplitude of the harmonics are represented respectively by the narrow peaks representing the mathematical maxima in the graph. The x-axis represents frequency of the constituent harmonic components in Hertz. The vertical axis represents the amplitude of the harmonics in dB. The harmonic structure of a vowel has characteristic hills representing the resonances caused by the different cavities primarily two major ones created by the height and front/back position of the tongue hump. These hills can be easily visualized if an envelope (thick line in Fig. 1.7) drawn covering the harmonic components. These resonances are commonly known as formants. As have been reported in the last section, the articulatory position of a vowel can be determined from the measurement of the formant frequencies.

In general first formant is associated with tongue height and the second formant frequency with the back to front position of the tongue hump (Ladefoged 1967). It is now common to use the first two formants for a reliable estimate for objectively determining the articulatory position of a vowel (Ladefoged 1967; Hartmut 2003). Figure 1.8 presents one example each for the seven Bangla vowels.

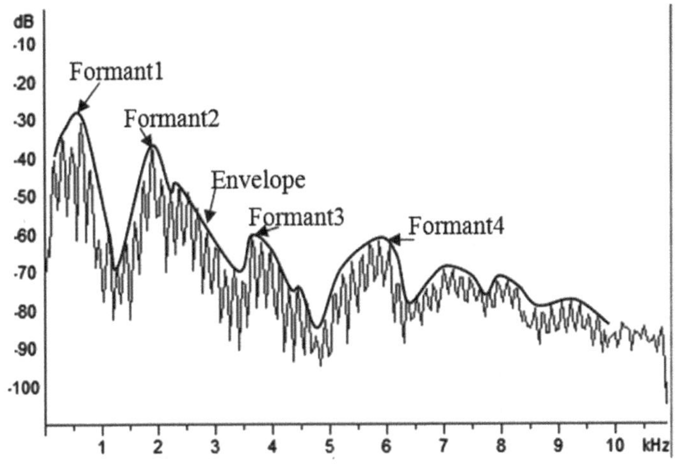

Fig. 1.7 Illustration of formants with respect to the vowel /æ/

For the determination of the articulatory positions of Bangla vowels the two-formant frequencies F_1 and F_2 were measured. Measurement of Formants Frequency was done from the steady state of the vowel segments. The segments were cut manually using wide band spectrographs. Only vowel segments with a minimum length of the steady state of 40 ms is considered for the study. This was done to assure that the vowel has reached its own quasi-stationary state.

The formants of the steady state vowel segments are then extracted automatically using Wave Surfer software (Wayland and Jongman 2003). The parameter settings for the extraction of the formants using Wave Surfer are as follows:

- Analysis window length 0.049 s
- Pre-emphasis factor 0.7
- Frame Interval 0.01 s.

The formant values of those vowel segments, which are outside mean ± standard deviation, are once again checked and if the formant extraction is found erroneous these are corrected manually using spectrum section. If the error is due to the contextual effect in cognition as indicated towards the end of Sect. 1.1 the data is rejected. For investigating the lip roundedness the video-frames during the steady state portion of the vowel was selected. Altogether 700 frames were analyzed. One may note here that the vowels in this dataset represent variations of each vowel sound because of the different contextual effects as well as the random variations associated with all recurrent natural bio-phenomena. All possible contexts are envisaged. It is known that some of the uttered vowels if cut out and listened in isolation might have really encroached into an adjoining category. This was rejected through the screening procedure mentioned above. It is expected that the choice of large number of possible context would allow such perturbation to be unbiased in directions and therefore the mean would give a fair representation.

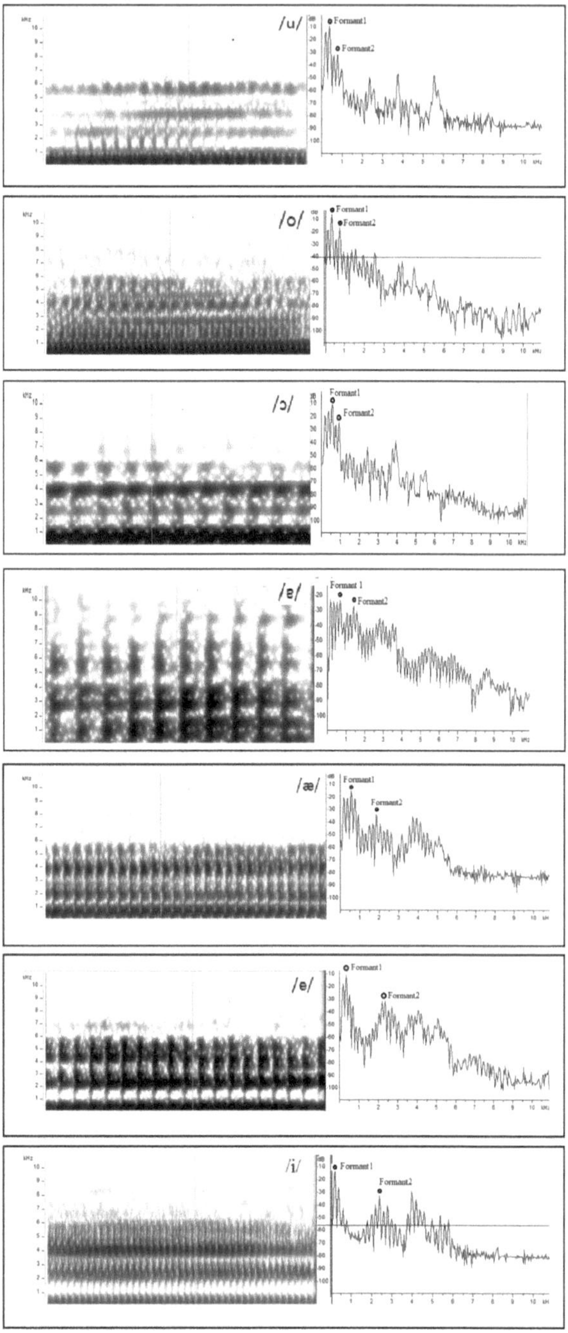

Fig. 1.8 Examples of spectrograms (*left*) and spectral sections (*right*) for each of seven Bangla vowels

1.3.2 Results

Table 1.3 presents the mean values and standard deviations (SD) of frequencies of F_1, F_2 and F_3 separately for male and female informants along with those for data pooled for both the sexes so that the difference due to sex, if any, can be clearly visualized. The SD is found to be always less than 25% of the mean. This is generally taken to be symptomatic of a good data set in speech research.

Figures 1.9, 1.10, 1.11, 1.12, 1.13, 1.14 and 1.15 show 3-D representations of the frequency distribution of seven Bangla Vowels in F_1-F_2 plane of male and female informants so that the difference due to sex, if any, can be clearly visualized. A careful perusal of these distributions reveal a clear separation in the distribution for male and female speakers for vowel /æ, ɐ/. The small secondary peaks seen in the distributions are some artifacts not separate peaks for the two sexes.

Figure 1.16 shows the sex-wise position of the vowels in the F_1-F_2 plane. The figure speaks of itself. Except for the two extreme back vowels /u, o/ all vowels for male speakers are pulled low in both F_1 and F_2 axes. Also interestingly male /o/ was nudging against female /u/. This does not seem to be a happy situation to which one had to live through more than half a century in presenting vowels in F_1, F_2 in an objective manner. However we shall see later in this section that taking account of the fundamental frequency along with F_1 and F_2 helps to resolve this problem.

Figure 1.17 represent the mean position and idea of the spread of each of the seven Bangla oral vowels in F_1-F_2 plane for data of both sexes pooled together. The dots represent the mean position of the vowels. The ovals give an idea of the spread where the widths and the heights of the ovals are standard deviations of F_2 and F_1 values respectively. Assuming normal distribution the ovals cover only about 68%

Table 1.3 Mean and standard deviations of F_1, F_2 and F_3 for all Bangla vowels

Vowel		Female			Male			Pooled		
		F_1	F_2	F_3	F_1	F_2	F_3	F_1	F_2	F_3
u	Mean	349	1026	2995	311	1023	2536	333	1029	2800
	SD	51.0	225.0	258.7	27.5	147.0	194.8	46.6	193.1	325.8
o	Mean	469	1112	2958	360	1041	2534	422	1094	2773
	SD	76.4	211.8	226.7	34.8	121.5	171.2	82.0	229.6	293.4
ɔ	Mean	665	1225	2862	480	1059	2489	582	1142	2695
	SD	136.5	235.3	208.6	75.4	83.8	221.4	145.9	215.1	283.4
a	Mean	899	1598	2819	673	1312	2503	798	1470	2679
	SD	149.5	198.2	277.3	131.5	144.2	248.7	180.9	226.3	308.0
æ	Mean	748	2023	2877	559	1775	2477	661	1909	2694
	SD	125.4	157.7	255.0	102.9	154.7	155.0	148.9	199.2	293.4
e	Mean	452	2383	3085	372	1972	2534	417	2204	2845
	SD	68.1	180.3	195.4	34.8	146.7	169.0	68.6	262.8	329.5
i	Mean	335	2613	3172	305	2198	2636	322	2430	2936
	SD	47.3	239.2	243.1	28.5	132.0	181.2	42.9	286.5	181.2

Fig. 1.9 Frequency distribution for vowel /u/

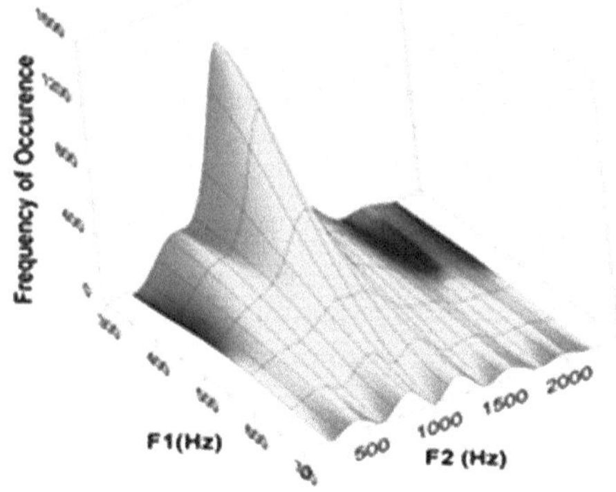

Fig. 1.10 Frequency distribution for vowel /o/

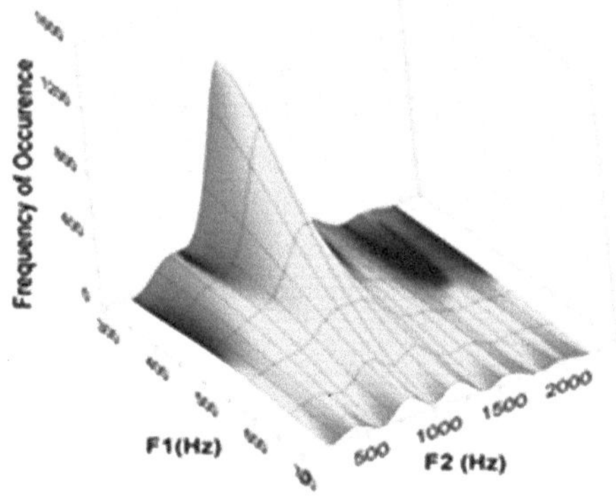

Fig. 1.11 Frequency distribution for vowel /ɔ/

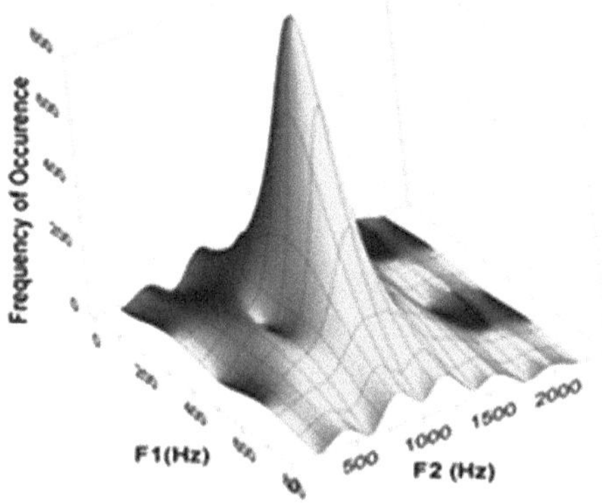

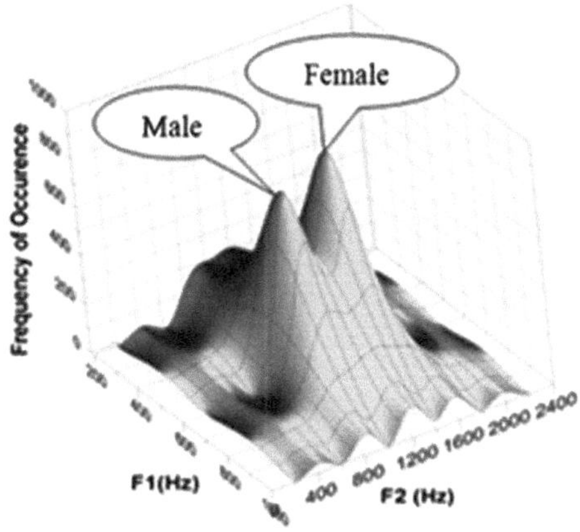

Fig. 1.12 Frequency distribution for vowel /a/

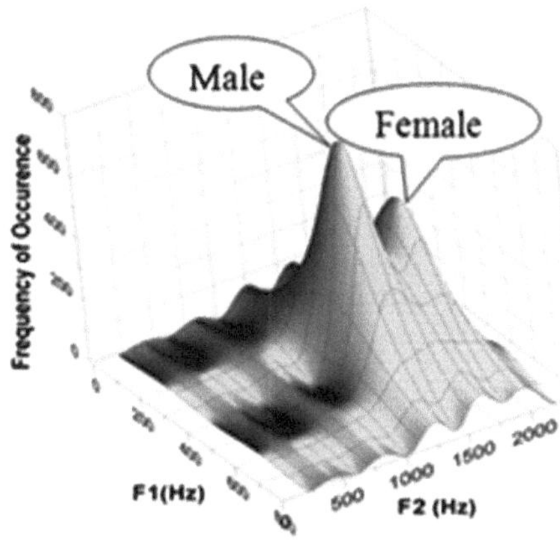

Fig. 1.13 Frequency distribution for vowel /æ/

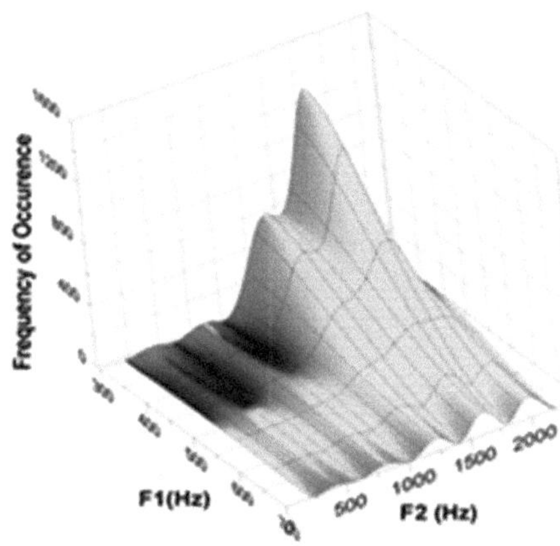

Fig. 1.14 Frequency distribution for vowel /e/

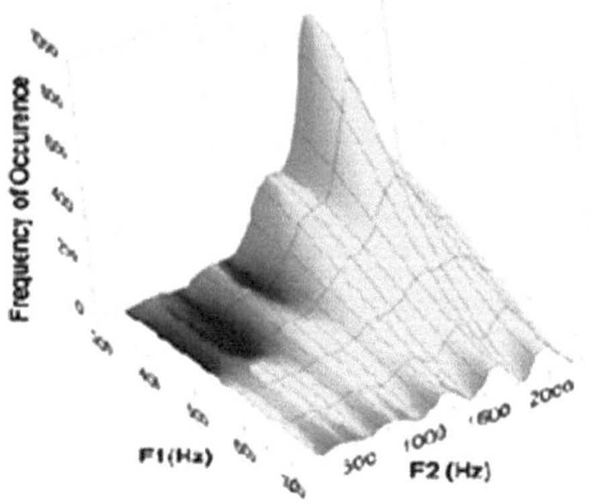

Fig. 1.15 Frequency distribution for vowel /i/

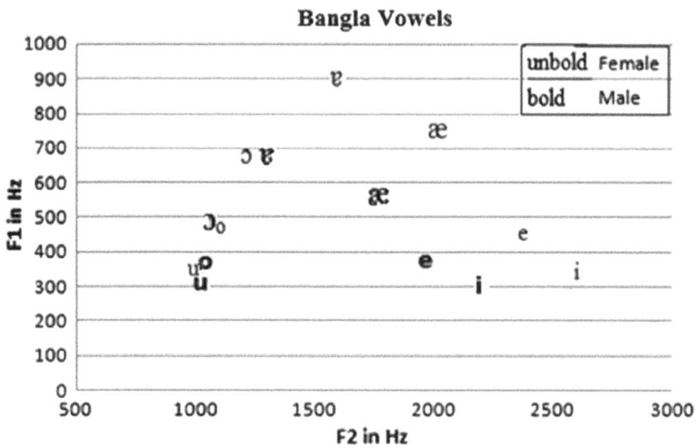

Fig. 1.16 Vowel position separately for male and female informants in F_1, F_2 plane

Fig. 1.17 Distribution of Bangla vowels in F_1-F_2 plane

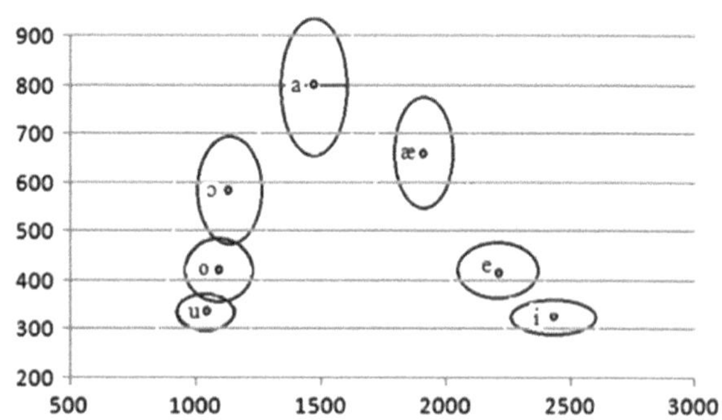

of the data. That the formant frequencies F_1, F_2, and F_3 closely follow normal distribution was reported as early as 1978 (Datta et al. 1978). Though the ovals appear to be disjoint actually this is so because they contain only a part of the data. In reality there is considerable overlap between the vowel clusters. Though the height and the forwardness of the hump of the tongue in the vowel diagrams presented normally show quite isolated positions, perceptual positions also overlap significantly in the normal vowel diagrams when actual field utterances are taken into account.

It may be understood that both production and cognition of vowels are not as crisp as we see in a vowel diagram. Men, women of different ages, and children possess vocal tracts of different shapes and sizes. The resulting formant frequencies are also different. Yet they produce vowel with equivalent qualities and people can decode them without much problem. This is because of the self organizing property of the cybernetics of speech evolution of an individual. Of course normally the context helps them. The "high degree of agreement among the judgments of skilled phoneticians" noticed by Ladefoged (1967) could be because of the fact that he used monosyllabic words. Dioubina and Pfitzinger found out that even phonetically trained subjects do not perfectly agree when judging vowel quality of isolated vowels.

A recently reported technique (Hartmut 2003) enables one to represent formant data, together with F_0 values, into the traditional perceptual evaluation of the category of a vowel utterance in terms of height and backness of the tongue. Formant data from which Table 1.3 is derived along with the value of fundamental frequency F_0 is used for the converting them into perceptual parameters of tongue height (h) and backness (b) using the following equations:

$$h = 2.621\log(F_0) - 9.031\log(F_1) + 47.873$$
$$b = -0.486\log(F_0) + 1.743\log(F_1) - 8.385\log(F_2) + 59.214$$

Mean and standard deviations of these parameters for each vowel from the data pooled for all informants are used to draw the diagram shown in Fig. 1.18. Diamonds represent the mean position of the vowels and ovals represent spread of the clusters.

It is interesting to note that the overlaps between the clusters have increased in the perceptual diagram in comparison to that in F_1-F_2 plane. It indicates that in machine recognition F_1, F_2 parameters could be more effective. This diagram also indicates that there may be a reason for a relook to fix the IPA symbols for Bangla vowels. This shall be taken up in Sect. 1.3.3 later.

For assessment of lip-rounding one has to take recourse to the video-graphic evidences. As mentioned earlier video of all speakers were taken while recording the speech signals. For assessing the rounding of lips only those frames were picked out where the lips are clearly depicted. All such clear frames for a vowel are visually examined for arriving at a final decision. Figure 1.19 presents the examples of such frames for the seven vowels. After analyzing all the selected frames for

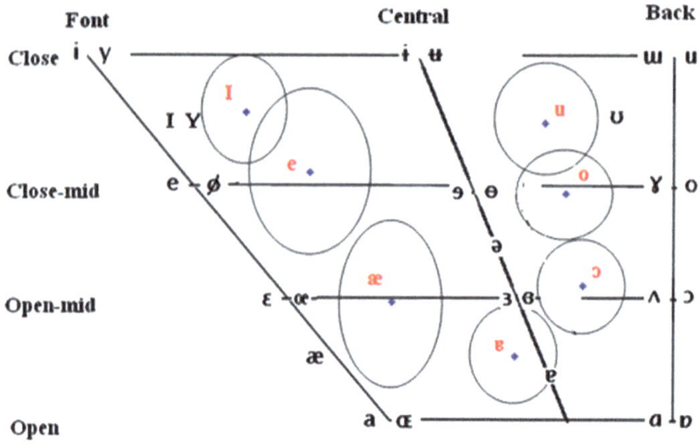

Fig. 1.18 Perceptual vowel diagram for Bangla vowels drawn from objective data

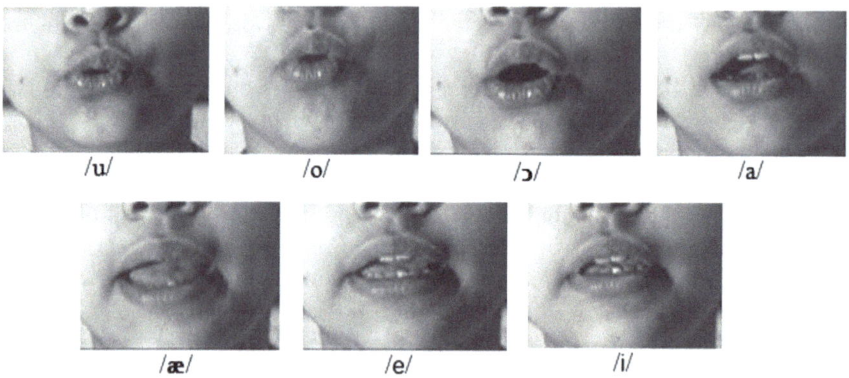

Fig. 1.19 Lip rounding for different Bangla vowel

production it is observed that in case of vowels /u/, /o/ and /ɔ/ lips are rounded. The degree of roundness is less in case of /ɔ/ compared to /u/ and /o/. It may be mentioned that it is possible to get objective quantitative measures of roundedness through computer processing of the images. It was not felt needed to do so for the simple reason that in vowel diagrams rounding is a binary decision.

1.3.3 Summary for Vowels with Native Read Speech

From the Cardinal vowel diagram (Figs. 1.16, 1.17 and 1.18) obtained through objective analysis of speech of a selected set of native speakers of both sexes the following points emerge:

Table 1.4 Traditional and actual symbols of Bangla vowels

	Traditional symbols for Bangla vowels (c/o previous literature)	Actual symbols	Description
1.	[u]	[u]	Close, rounded back vowel
2.	[o]	[o]	Close-mid, rounded back vowel
3.	[ɔ]	[ɔ]	Open-mid, rounded back vowel
4.	[a]	[ɐ]	Open, unrounded mid vowel
5.	[æ]	[ɛ]	Open-mid, unrounded front vowel
6.	[e]	[e]	Close-mid, unrounded front vowel
7.	[i]	[ɪ]	Close, unrounded front vowel

- The height of the Bangla vowel which is represented by **[u]** in the IPA chart is much lower than that of the cardinal vowel [u] and closer to Cardinal vowel **[ʊ]**.
- The height of the Bangla vowel **[æ]** is quite close to the cardinal vowel [œ].
- Bangla neutral vowel lies almost midway between [ɘ] and [ɔ] and since it is a back vowel one may choose [ɔ].
- The height of the Bangla vowel **[i]** is lower than that of the cardinal vowel **[i]** and closer to **[I]**.
- The Bangla vowel which is represented generally by the IPA symbol **[a]** is much closer to the cardinal vowel **[ɐ]**.
- All Bangla vowels are quite centralized in comparison with the cardinal vowels.

Table 1.4 gives the tentative IPA symbols dictated by Fig. 1.18 along with the traditional symbols.

1.3.4 Vowels in Controlled Environment

While the aforesaid studies use data from sense words used in continuous speech the vowels are not necessarily always properly pronounced. While the change of vowels in different contextual context is expected to differ in continuous speech the errors usually go beyond that. It is true that for technology development as well as to understand to what extent vowels get perturbed in actual speech the above results are very important. However at the same time it is necessary to know the position of vowels when they are spoken with attention. Let us call this 'formal pronunciation' as against the 'informal pronunciations' of spoken sentences presented in the previous sections. For studying formal pronunciation the same set of informants were asked to utter nonsense words of the form /cvcvcvcv/. Being aware that personal bias may still creep in the utterances were subjected to a listening test by three senior Bangla linguists for goodness of pronunciation. The steady states of the vowels from the selected proper pronunciations are used for extraction of formants. Table 1.5 gives the necessary statistics for F_1 and F_2 for male, female informants as

Table 1.5 Mean and S. D. for F1, F2 and F3 for 'proper pronunciation' of Bangla vowels

		Female				Male				Pooled			
		Formant 1		Formant 2		Formant 1		Formant 2		Formant 1		Formant 2	
		Freq.	N	Freq.	N	Freq.	N	Freq.	N	Freq.	N	Freq.	N
u	Mean	339	29	927	32	316	53	803	43	324	82	1005	89
	SD	20		134		36		174		33		594	
o	Mean	472	24	975	27	368	48	808	46	403	72	890	74
	SD	54		103		19		112		60		221	
ɔ	Mean	756	30	1116	33	538	40	906	42	622	75	1047	78
	SD	151		125		42		156		152		302	
a	Mean	963	38	1431	30	736	47	1187	45	828	85	1296	77
	SD	126		211		78		80		172		207	
æ	Mean	871	24	2289	25	578	44	2089	45	689	69	2159	69
	SD	120		98		54		190		164		191	
e	Mean	421	28	2645	33	358	47	2206	47	382	75	1540	80
	SD	72		200		26		77		57		872	
i	Mean	333	28	2833	28	291	44	2279	46	307	64	2511	67
	SD	16		149		20		94		31		300	

well as those for data pooled over the sexes. In this table N represents the number of samples. The small discrepancies that are observed between N for F_1 and F_2 is because some data is rejected if not found clearly from spectrograms.

A comparative perusal of Tables 1.3 and 1.5 reveals that F_1 of [a] and [æ] is lowered in conversational speech indicating that the jaw is not lowered fully. This is quite expected in free speech as the extra effort and attention required for proper pronunciation of these vowels may not be maintained in normal speech. For other vowels the difference in F_1 is not significant. One would expect neutralization of F_2 for conversational speech compared to proper pronunciation. It is corroborated generally since F_2 is raised for back vowels and lowered for front vowels in conversational speech. The only exceptions are [ɔ] and [e].

One should expect that the spread of the data distribution to be less for controlled pronunciation. One way to check this is to look at the standard deviations. While doing this one should take a comparative measure of SD with respect to mean values. The average of the ratio SD/Mean for all data for both the formal pronunciation and the informal pronunciation was calculated. The result came out as 0.145399 and 0.146205 respectively for informal and formal pronunciation, surprisingly almost the same value. The reason is not difficult to rationalize. It only indicates that the contextual influences on the parameters much outweigh the inadvertent errors in speech production. It only strengthens the notion of efficacy in the psycho-biological cybernetics involved in human speech development.

1.4 Nasal Vowels

In SCB all the seven oral vowels is known to have their nasal phonemic counterpart. Since nasalization of vowels are phonemic in Bangla it is expected that there would be some acoustic cue or cues consistent enough for perceptual distinction. It is therefore necessary to give special attention to this aspect for Bangla. Nasal vowels are produced when the velum is open and the nasopharynx is coupled with the oropharynx. If one examined Figs. 1.3 and 1.4 one may note that depending on the position of tongue hump the nasal cavity can act as a shunt to the first cavity or both the front and back cavity. When the nasal cavity acts as a shunt its resonances may act as the zeroes for the system causing antiformants to appear. Otherwise they introduce further formants sometimes as separate ones and at other times strengthening the oral resonances. This is a gross simplification of a complex process. However it helps to have an idea how the coupling of nasopharynx changes the timbral quality of produced sounds.

Some such cues are reported for some of the western and other foreign languages (Fujimara 1960; Berkins and Stevens 1982; Takeuchi et al. 1975; Hawkins and Stenens 1985). In general these studies reveal following acoustic cues for distinction of nasal over oral:

A. Strengthening of F_0
B. Weakening of F_1
C. Strengthening of F_2
D. Raising of F_1 and F_2
E. Presence of nasal formants and anti formants.

As regards to the cues A to D studies in SCB reported (Datta et al. 1998) that the strengthening of F_0 on nasalization is observed for all central and front vowels except [ɛ̃], the weakening of F_1 for all except for vowel [ĩ] and [ɔ̃] and the raising of F_2 except for vowels [õ] and [ũ]. This is mostly because of the nasal antiformants introduced due to the addition of nasal cavity in the resonating system through the opening of the velum (Hawkins and Stenens 1985). Examination of spectrograms shows consistent occurrences of nasal formants. For all vowels taken together nasal formants are found to be clustered in the region of 900, 1250 and 1650 Hz. For [ẽ] and [ũ] these are found between F_1 and F_2. Nasal formants are found stronger for male informants. These are generally stronger for front vowels. Only exception is the occurrence of an additional formant in oral [i].

Antiformants mostly lie below 1000 Hz. Majority of them lie within the range of 400–600 Hz. Except for [ũ] and [ĩ] nasal antiformants mostly lie within the range 200–600 Hz. Those for [ũ] are distributed over a wide range 1–1.9 kHz. For vowel [ĩ] no nasal antiformants was observed. On the other hand antiformants were observed for oral [i] in the range 800 Hz–1 kHz. For all vowels except [ũ] and [ĩ] at least one antiformant is observed below 1 kHz. Both the places and the occurrences of nasal formants and anti-formants in the referred case study in Bangla has been found to lack sufficient consistency regarding distinction between the nasal vowels themselves though they are quite indicative of nasalization.

The multiplicity, complexity and somewhat fluid situation regarding acoustic signature of nasality in vowel from the standpoint of machine recognition led to a study for the search of a single necessary and sufficient acoustic cue for oral/nasal distinction in Bangla (Datta et al. 1998). The study used analysis—through—synthesis paradigm and reported that such a cue do exist in the case of Bangla. This was one or two harmonics between F_0 and F_1 lying in the neighborhood of 400 Hz. Figure 1.20 gives the example showing the harmonics which plays a pivotal role in the nasal/oral distinction. The upper part of the figures shows the spectrum section at the first vowel. The second harmonics is responsible for distinction. This is indicated by the crossing of curser lines. It may be seen that for vowel /ɐ/ this is about 12 db higher than the corresponding nasalized /ɐ̃/. It has been noticed that for high vowels /u/ and /i/ they are found to increase for nasal counterparts. For other vowels they decrease. The very consistent nature of this cue calls for a detail investigation into the relationship of the glottal waveform and the opening of velum. Such a relationship has been observed in the case of aspirated vowels as we shall see in a later section.

We have already noticed that the introduction of nasal formants causes, inter alia, shifting of the first two formants namely, F_1 and F_2. It is therefore necessary to visualize nasal counterpart of each oral vowel in the F_1-F_2 plane.

Figure 1.21 presents the mean values of the first two formant frequencies of oral along with its nasal counterparts of the seven vowels in SCB. The difference of first formant frequencies between oral and nasal vowel is small, in most cases insignificant, except for the mid vowels /e – ẽ/ and /o – õ/. The second formant is decreased on nasalization for two back vowels /o/ and /u/, however it is found to

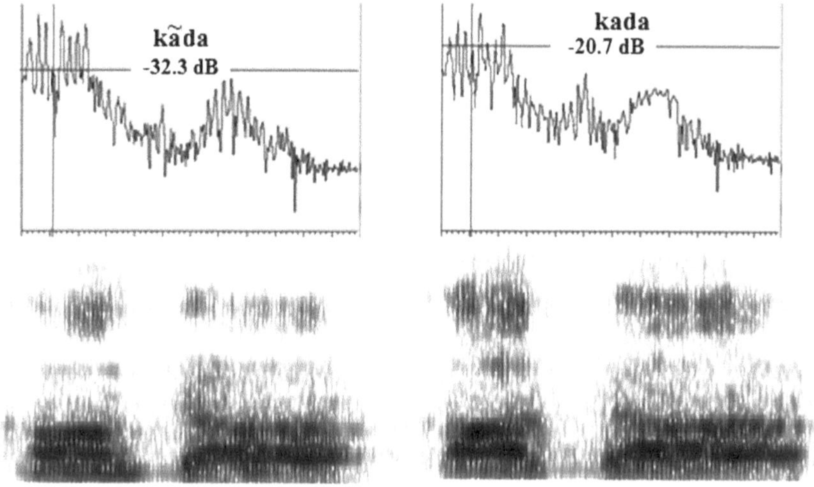

Fig. 1.20 The harmonic responsible for nasal/oral distinction

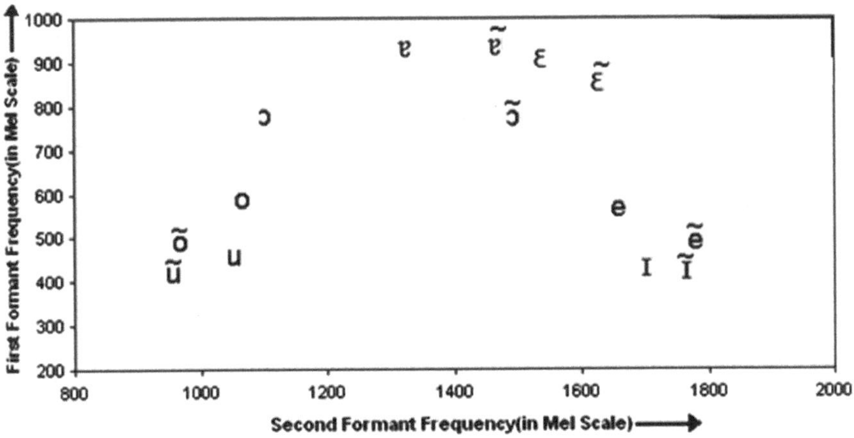

Fig. 1.21 Mean frequency distribution of seven Bangla of both oral and nasal vowel

increase extraordinarily for the neutral vowel /ɔ/. For the other vowels there is significant increase of the second formant frequencies on nasalization. These changes in the formant structure may be due to the intervention of formants and antiformants introduced by the coupling of the nasopharynx as a result of opening of velum.

1.5 Aspirated Vowels

Murmured and whispered vowels fall under the category of aspirated vowels as the vocal chords are held apart further than in the case of normal or clear vowels. A larger volume of air escapes between them producing an audible noise normally referred to as aspiration. The existence of murmured and whispered vowels is reported in various languages including some Indian languages. In Gujarati eight murmured vowels as well as eight murmured nasalized vowels has been reported in contrast with corresponding clear oral and nasal vowels. Murmured vowels in Marathi also have been reported (Masica 1991). Whispered vowels have been noticed in the western dialect of Awadi (Saksena 1971). The objectivity of these sounds has been widely researched in terms of finding the acoustic contrastive cues. However, historically these vowels are regarded as allophones of post-vocalic /h/ (Masica 1991). Esposito et al. (2007) reported breathy phonation in /Nh/ clusters and some nasals in some Indian languages including Bengali. The presence of the character [h] in some of the Indian scripts is associated with the presence of murmured vowels in the corresponding spoken form in many Indian languages (Wayland and Jongman 2003; Mistry 1997; Pandit 1954; Masica 1991). Apparently for Standard Colloquial Bengali (SCB) there has been, in general, a lack of interest to investigate the objective as well as subjective existence of this phenomenon. The whole attention seems to have been devoted to characterize phonation corresponding to the grapheme হ ([h]) acoustically in terms of whether they are voiced or unvoiced along with the presence of some turbulent glottal noise. As we shall see later that in SCB there exist a wide range of variance in spectral structures in various instances of the so-called [h]. We shall also see that these could be reasonably grouped into some specific patterns close to but distinctly different from those observed in clean vowels. It stands to reason to say that a phoneme in any language must have a characteristic coherent spectral structure including those reflected in the resonances (formants) and anti-resonances (antiformants) determined by the articulatory configuration. This section intends to address this issue.

Taking cue from the aforesaid Indian languages an attempt is made to find murmured and whispered vowels, if any, in standard colloquial Bengali. The evidences from other referred Indian languages allowed the area of search to be narrowed down to investigate the acoustics of the signal in the neighbourhood of /h/ in such words. Hence, 37 words containing /h/ were selected with the expectation of locating murmured and whispered vowels in standard colloquial Bengali. These words spoken by 5 male and 5 female speakers in a neutral carrier sentence were taken from the speech data base of CDAC, Kolkata as the data for the study. Standard soft ware packages namely Cool-Edit Pro and Wave Surfer was used for the extraction of acoustic parameters.

The acoustic signatures also have been investigated in detail elsewhere (Fischer-Jorgensen 1967; Thongkum 1988; Wayland and Jongman 2003; Andruski and Ratliff 2000; Hombert et al. 1979; Pandit 1954; Hillenbrand et al. 1994; Hillenbrand and Houde 1996; Klatt and Klatt 1990). A synopsis of acoustic parameters normally studied to characterize murmured and whispered vowels as against clean vowels seems to be in order. The extensively investigated five parameters are: (a) the ratio of energy in low to high frequency bands, (b) the ratio of total energy of the candidate with respect to that of the adjoining clear vowel, (c) the ratio of the fundamental frequency of murmured to adjoining clear vowels, (d) the lower formant frequencies and (e) the relative amplitude of the first two harmonics. These along with a new parameter namely, the slope of periodic decay associated with a period of a vowel in the signal has been investigated. This parameter has been explained in detail in the next section.

The experimental results on SCB strongly indicate the presence of murmured and whispered counterparts for almost all the clear vowels of the dialect. All these are found to represent the character [h] in the textual representation of the word. It is seen that only 7 out of the 257 samples examined were found to be pure fricative noise.

1.5.1 Acoustics of Aspirated Vowels

Murmured voice (also called Breathy voice) is a phonation in which the vocal cords vibrate, as they do in normal (modal) voicing, but are held further apart, so that a larger volume of non-pulsating steady air also escapes between them (Laver 1980; Gordon 2001; Chávez-Peón 2013). This produces an audible noise normally referred to as aspiration. The vocal cords while held slightly apart are lax so that they vibrate loosely. Sometimes the vocal cords are brought closer together along their entire length but not as close as in modally voiced sounds such as clear vowels. This results in an airflow intermediate between a fricative and vowels. Another way to produce a murmured vowel is to constrict the glottis, but separate the Arytenoid cartilages that control one end. This results in the vocal cords being drawn together for voicing in the back, but separated enough to allow the passage of large volumes of air for voicing in the front. The produced acoustic features are quite distinctive from those of clean vowels on one hand and the pure sibilants on the other hand. In general the glottal gesture in producing aspirated vowels is such that it allows a continuous stream of air to flow along with oscillations of vocal folds.

The whisper has been regarded occasionally as a simple modification of breathy voice. Whispered speech is produced by modulating the flow of air through partially open vocal folds. Because the source of excitation is turbulent air flow, the acoustic characteristics of whispered speech differs from voiced speech. Since the aspiration source is located near the glottis, no zero appears in the trans-conductance between this series pressure source and the acoustic flow at the lips (Ngoc and Badin 1994). Physiological measurement of the laryngeal shape during whispering by using magnetic resonance imaging shows that the supra-glottal structures were not only constricted but also shifted downward, attaching to the vocal fold to prevent vocal fold vibration (Tsunoda et al. 1997). Acoustic analyses show that for whispered vowels the 'formant-like' features (i.e. the hills in the spectral structure) in the region of F1, F2, and the global peaks have a tendency of upward shift. These are also much flatter than those for clear vowels (Tartter 1989). The spectral tilts are also flattened compared to the clean vowels. The intended pitch by a talker does not correspond to a specific formant frequency (Konno et al. 2006).

When the air stream coming out through the glottis is not fully pulsed there is a portion which is a steady flow. If the passage is narrow and the velocity is strong enough turbulence is created. This is the aspiration noise component additional to the turbulence created by the oscillation of mucosal cover over the rigid muscles of the vocal chords. The later turbulence is associated also with normal vowels in the form of random perturbations namely jitter, shimmer and HNR/Complexity perturbation (Dutta et al. 2003). This additional noise is characteristic of murmured vowels (Klatt and Klatt 1990; Hillenbrand et al. 1994; Stevens 2000; Ladefoged and Antananzas-Barroso 1985) and may be assessed using a band-cut filter at about F3. In fact this noise may actually replace harmonic excitation of higher formants. The aspiration noise would be referred to as 'N' in the following sections.

In the production of murmured vowels there is an acoustic coupling with the trachea (Fant et al. 1972). The acoustic effects of tracheal coupling on the normal transfer function of the vocal tract for a vowel include possible addition of poles (formants) and zeros (antiformants) associated with the tracheal and lung systems. This causes loss of energy in the vocal tract (Stevens and Hanson 1994) due to the resistance of the yielding walls of the vocal tract, and heat conduction and frictional losses at the walls. Thus a comparison between the total energy of a murmured vowel with that of an adjoining clean vowel, if there is one, may be an effective indicator. Cross-linguistic investigations of phonation types generally show that breathy phonation is associated with a decrease in overall acoustic intensity in many languages including Gujarati (Fischer-Jorgensen, 1967; Thongkum 1988). This finding is, however, not universal. Wayland and Jongman (2003) for example found that breathy vowels in Javanese are associated with an increase in overall acoustic intensity. The ratio of energy of the steady state of murmur to that of the adjoining vowel will be denoted by 'E' henceforth.

During the production of breathy phonation, to allow the vocal folds to vibrate while they stay relatively far apart, the vocal folds have to be relatively less taut. Thus, the fundamental frequency of a breathy vowel is expected to be lower than that of a clear vowel. This expectation was borne out in Javanese and Green Mong (Wayland and Jongman 2003; Andruski and Ratliff 2000). This may also explain why breathy phonation appears to be consistently associated with lowered tone in many languages reviewed earlier (Hombert et al. 1979). It therefore seems reasonable to view the difference of the fundamental frequency of the murmur segment and the adjoining clean vowel as an indicator of murmur.

Not much information is available on the values of the first two formant frequencies for murmured vowels though measurements of F1 bandwidth is said to provide an indirect indication of murmurs (Stevens and Hanson 1994). However for the whispered vowel the lower formant frequencies are known to be slightly higher than those of the modal vowel (Pandit 1954). It therefore seems necessary to examine the first two formant frequencies for the aforesaid aspirated vowels in comparison to those of the adjoining clear vowels.

The round near-sinusoidal shape of breathy glottal waveforms is responsible for a relatively high amplitude of the first harmonic (H1) and relatively weak upper harmonics. However, in order to assess whether there is an increase in the H1 amplitude it must be compared with some reference. The amplitude of H2 for this purpose has been suggested (Hillenbrand et al. 1994; Hillenbrand and Houde 1996; Bickley 1982). Hanson (1995) reported that for breathy phonation the glottal configuration is adjusted in such a way that a larger open quotient results, while rate of decrease of airflow at glottal closure remains nearly the same allowing an increase in the difference between H1 and H2. A spectral analysis of H1 amplitude relative to that of H2 for !Xoo and Gujarati vowels by Bickley (1982) revealed that the amplitude of H1 was higher than the amplitude of the adjacent H2. Klatt and Klatt (1990) noticed H1/H2 amplitude differences between naturally produced breathy and clear word pairs is around 6 dB for Gujarati and 9.7 dB for !Xoo data.

Enhanced H1 amplitude in the spectra of breathy voice signals has been observed by a number of investigators.

Vowels are produced by glottal pressure pulses finally shaped by the resonance of the supra-glottal cavities. This is a sort of repetitive forced vibration of a resonating system. The shape of the resulting signal is determined by the shape of the forcing pressure pulse at the beginning and when the pulse dies out the resonating system takes control. During this time the signal decays at a rate determined primarily by the damping factor of the resonating system if it is isolated from any source of energy. This is so for clear vowels when glottis is fully closed. However for murmurs the glottis is never fully closed and therefore the coupling of the sub-glottal structure with oral cavity is likely to increase the damping. This increase in damping would cause the decay to be faster. On the other hand there is a possibility of deriving energy for the signal from the glottal air flow which would tend to slower the decay of the signal in a period. Thus the decay of the signal in the wave form in a period of vowel utterances can be an acoustic feature worth investigation for examining breathiness in a vowel pronunciation.

1.5.2 Methodology

The aspirated vowels are found to occur profusely when there is the character [h] in a word as reported (Mistry 1997; Pandit 1954; Masica 1991) for some other Indian languages. Therefore, for the present study 37 SCB words, embedded in a neutral carrier sentence, with /h/ in the syllable are taken from the speech corpus of CDAC-Kolkata. Of these, there are 21 words where /h/ is word-initial the rest are word -medial. The sentences were spoken by 7 male and 7 female informants in the age group of 19–45 years. Only 257 samples could be used for analysis after rejecting the instances of unclear pronunciations and short duration (<30 ms) of the steady states of /h/.

As already mentioned aspiration noise mainly introduces additional energy at the high frequency end and a standard way is to compare energy balance using F3 region as the boundary. For SCB the F2 rarely exceeds 2.8 kHz. Therefore, for the present study this is used as the boundary for spectral balance. For this purpose of filtering and energy measurement more or less the steady spectral region of the segment, murmur or adjoining clean vowel, is used. Figure 1.22 presents an example with the word /bɐɦɐdur/ where /ɦ/ is in word medial position. The portions inside the vertical lines indicate the nearly steady regions used for acoustic measurements.

A sharp band-cut filter with −30 dB attenuation with cut-off at 2.8 kHz has been used for measuring the high (h) and low (l) frequency energy content in a segment (Fig. 1.23). As mentioned earlier the amount of aspiration noise 'N' is represented by the ratio h/l expressed in dB.

As said before the difference of the signal energy between a clean vowel and the aspirated vowel that is another indicator of the breathy character. Fortunately in

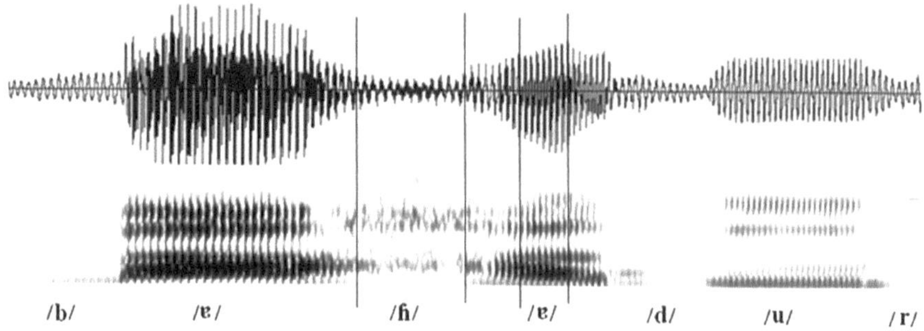

Fig. 1.22 Example showing region selected for acoustic measurements

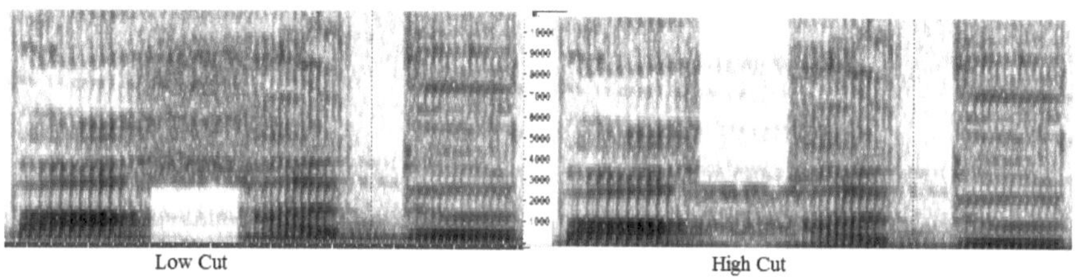

Fig. 1.23 Example showing results of filtering to deduce aspiration noise

almost all the cases studied here /ɦ/ has the colour of the succeeding clear vowel. Therefore, it is easier to select the required segments for total energy comparison (Fig. 1.23). The parameter ‘E’ is given by the ratio of the total energy of the selected segment of /ɦ/ to that of the adjoining vowel expressed in dB.

As mentioned earlier the difference between the fundamental frequencies of the murmur vowel segment and the adjoining clean vowel may be studied as a cue for murmur. This difference is defined here as $\Delta f_0 = ɦf_0 - vf_0$ where $ɦf_0$, vf_0 respectively denotes the average fundamental frequency of the steady states of [ɦ] segment and the adjoining vowel segment.

The amplitudes of the first and second harmonics (H1, H2) are measured from the spectrum section at the steady state of the voiced region (Fig. 1.24). For comparative assessment of these amplitude measures, the steady states for [ɦ] and the adjoining clean vowel segments were used.

The spectral tilt is a very general term and has been used for different regions in the spectra, e.g. this may represent amplitude of F_1 or F_2 with respect to the f_0 or even comparing H1 and H2. For the present study spectral tilt represent the tilt of the full spectra from f_0 to 10 kHz. It is measured by taking the spectrum section of the whole of the steady state and then the trend line for the whole spectra. Figure 1.25 shows the average spectral plot of the selected nearly steady state of the segments of three different types of vowel namely, a clean vowel, a murmured vowel and a whispered vowel selected from the present SCB data base. The individual plots are vertically shifted for the ease of comparison. One can easily

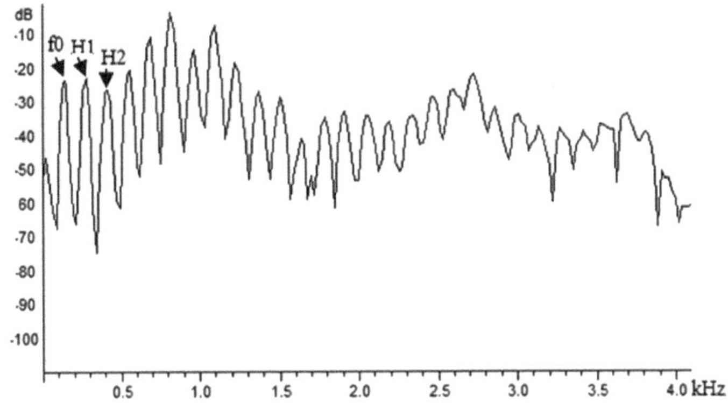

Fig. 1.24 Spectrum section of a normal vowel showing H1 and H2

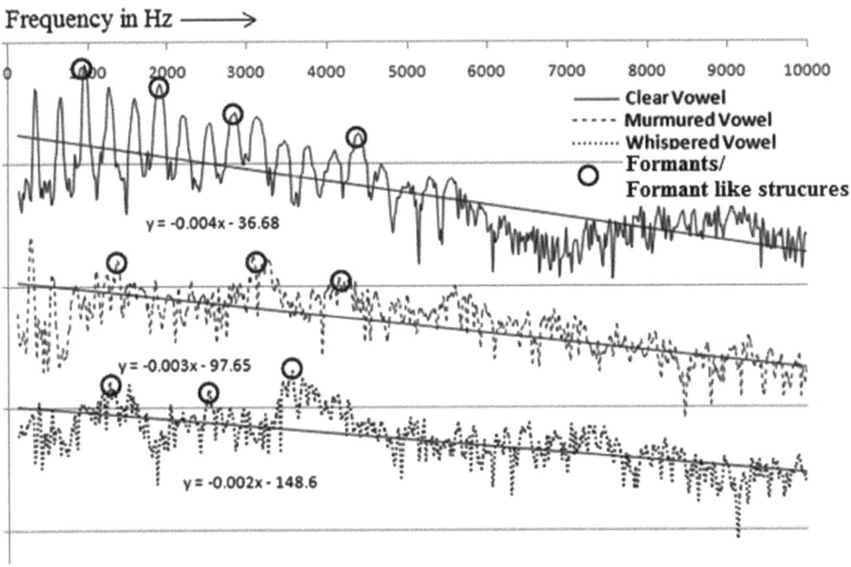

Fig. 1.25 Spectral structure of different types of vowels with the trend lines and slopes

visualize the differences. The equations relate to the corresponding trend lines. The coefficient of x gives the value of slope. This is used to represent spectral tilt for the present purpose. Formants and formant like structures are indicated by circles in the figure.

As vowels are produced by repetitive excitation of supra-glottal cavities by a series of pressure pulses coming from the glottis they exhibit a decaying periodic structure. In most cases it is not difficult to segment the signal into the periodic

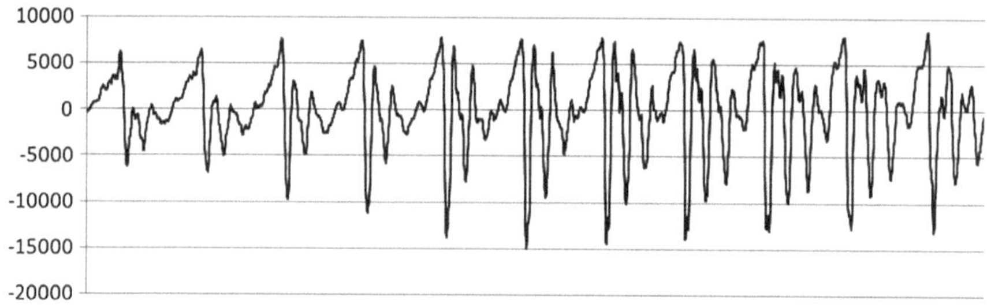

Fig. 1.26 Example of a clear vowel signal

structures. For the determination of periodic decay the following procedure is adopted:

1. The absolute value of the signal, which folds up the negative portions of the signal, is used for the purpose.
2. Each period is amplitude normalized at a pre-fixed value say 10,000.
3. For each period a line is drawn which just covers the signal.
4. The slope of the cover is taken as an estimate of the decay.
5. Relatively steady states of the vowels are used.

Figure 1.26 shows the clear vowel signal and Fig. 1.27 shows the corresponding signal with negative parts folded up and amplitude normalized for each period. The dashed lines show the cover of this transformed signal the slope of which is taken as the estimate of the decay factor for each period. The covers are drawn manually.

1.6 Results and Discussions

There are 257 usable signal files from 21 words with character [h] spoken by 14 informants of both sexes. Of all these only 7 (<3%) has been found to be pure sibilants. Clear vowels are characterized by a substantively unobstructed passage through the supra-glottal pathway as against consonants which are characterized either by substantive obstruction or constriction. These seven records are heard by the author as either murmured or whispered vowels. In fact these are heard as a variant of a vowel like sound with the same colour as that of the adjoining vowel, succeeding ones in most cases. In Bangla, [হ] generally manifests itself as different classes of spectra each representing closely one of the seven vowels. It seems contrived to consider phones having different spectral structures, which could be distinctly categorized in different categories and heard as sounds which are categorically different, as a single phoneme. In this context one needs to examine carefully the acoustic segments representing the character [হ], in SCB, in terms of the signatures known to characterize murmured vowels in other languages.

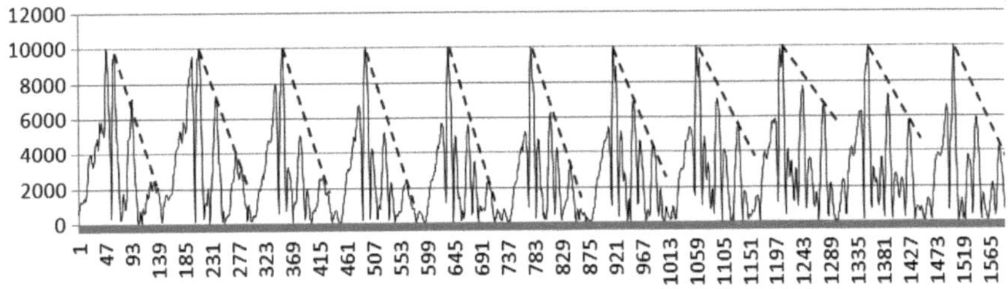

Fig. 1.27 Example showing folded normalized signal with the cover lines

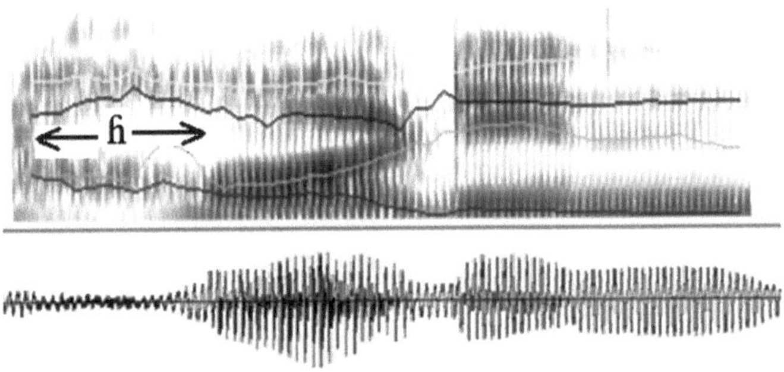

Fig. 1.28 Acoustic features of the word [ɦɔren]

Figures 1.28 and 1.29 presents examples of the acoustic picture of these sounds representing the [h]. All the four formants (represented by solid lines of different shade of grey) are clearly visible and consistent with those in the succeeding vowel regions. As we shall see later, there are certain acoustical structures which strongly favour [ɦ] to be considered in the group of murmured vowels, as has been reported for some other Indian languages mentioned in Sect. 1.5, instead of glottal fricative. These are presence of strong fundamentals and distinct formant structures which closely correlate with the succeeding vowel. The waveforms appear to be similar to vowels (mostly the following ones) with some high frequency perturbations added and amplitude lowered.

Figure 1.30 presents the frequency distributions of the aspiration noise N for the [ɧ] segments and the adjoining clear vowels. Aspiration noise is seen to be clearly higher for [ɦ] segments. The modes are well separated. The cross-over point is at −12 dB. The fact that with this boundary, 225 (90%) out of 249 adjoining vowel segments, fall in the category of clear vowels clearly indicates that N is a good indicator of murmurs. Again with this boundary 182 out of 249 [ɧ] segments (73%) fall out of the category of clear vowels. These may be considered as aspirated vowels which, in isolation, are heard by the author as either murmur or whisper. Of the 20 [ɧ] segments for which N is positive, which means that the energy content

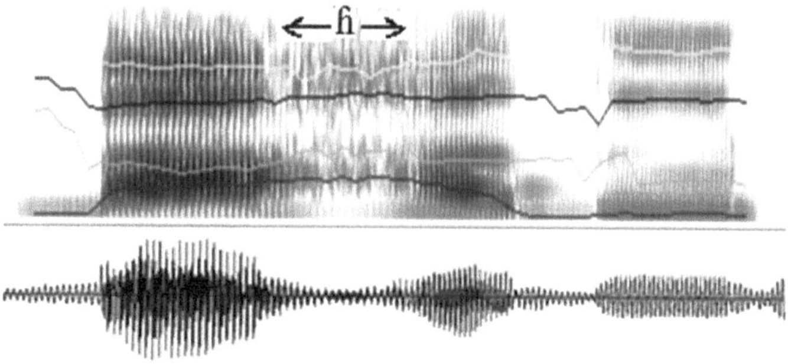

Fig. 1.29 Acoustic features of the word [bɐɦɐdur]

Fig. 1.30 Distribution of the aspiration noise N for clean and murmured vowels

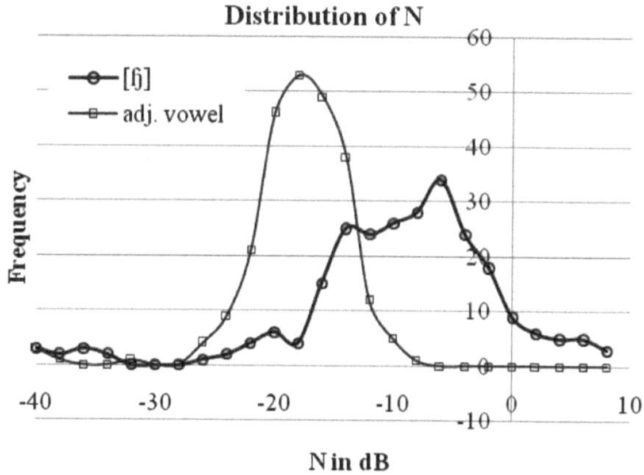

above 2.8 kHz is more than that below, one is murmur, two are sibilants and the rest are whispers.

Average power of a murmur with respect to that of the adjoining vowel is reported (see Sect. 1.5.2) as a cue for murmur. Figure 1.31 shows the distribution of E, the ratio of the average energy of the steady state of murmured segment to that of the adjoining vowel. All murmurs are found to be weaker with only one exception. The mode occurs at −20 dB indicating this cue to be quite robust.

Figure 1.32 presents the distributions of differences H1-H2 for the murmur segment and the adjoining vowel segment. For SCB the two distributions appear to be equivalent. Both are symmetric with the mode near about 0 (3–6 dB) looks like a normal distribution. The standard deviation for both of the distribution is about 10.5 dB and is reasonable. Thus unlike Gujarati and !Xoo (see Sect. 1.5.1) H1 and H2 does not show discriminatory property in case of SCB murmur against clear vowels.

Figure 1.33 shows the distribution of differences in f_0 (adjoining vowel − murmur). The f_0 is lower for murmured vowel except only for 14 samples,

Fig. 1.31 Comparison of total energy of the murmured segment wrt that of adjoining vowel

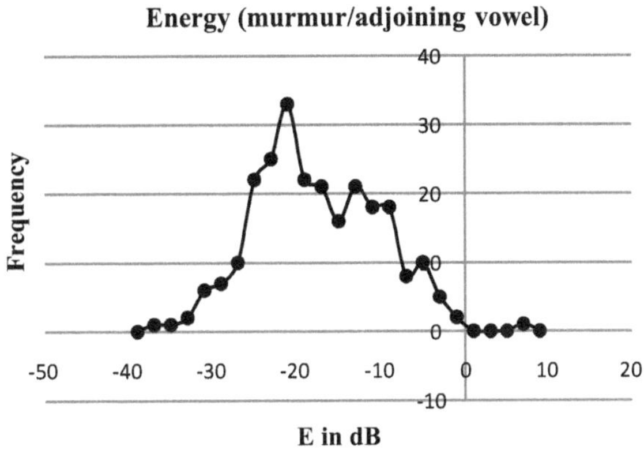

Fig. 1.32 Comparison of H1 and H2 for murmured segment and adjoining vowel

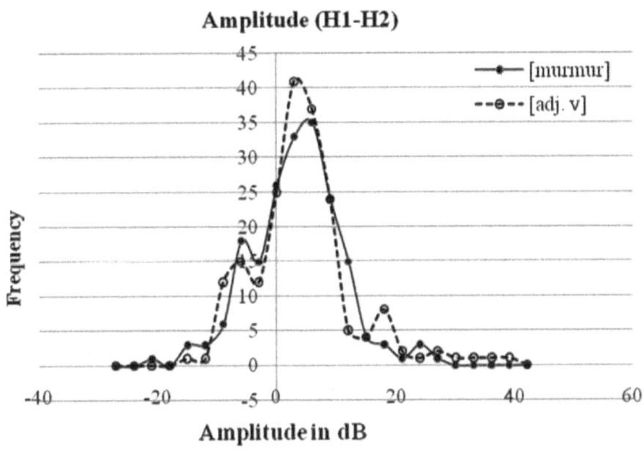

Fig. 1.33 Distribution of differences in fundamental frequency

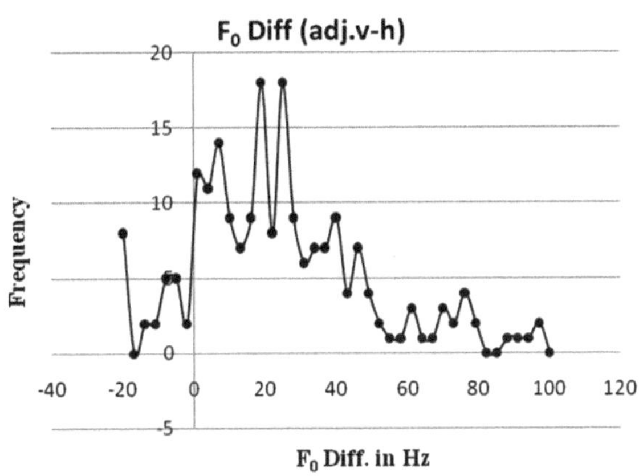

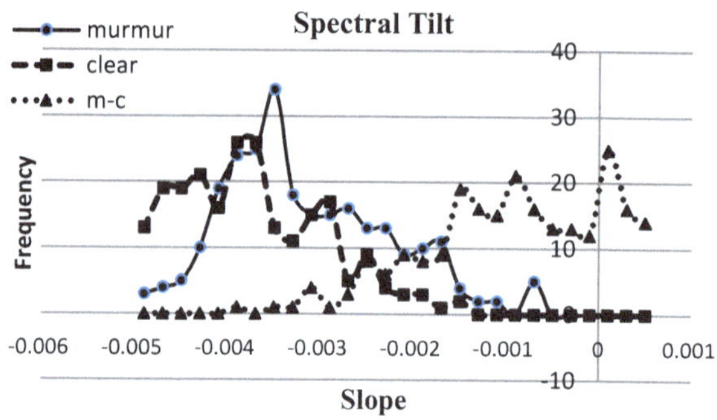

Fig. 1.34 Distribution of spectral tilts of murmurs, adjoining vowels and the differences

which may be considered as negligible. The distribution is slightly skewed with the mode around 23 Hz. This is about 10% of the average value of f_0 of the clear vowels in the sample set. The distribution is drawn with 208 pairs. For the rest murmur segments f_0 could not be ascertained with confidence. Thus the expectation that murmurs have lowered f_0 is consistently vindicated though the average increase is not very high.

Figure 1.34 presents the frequency distribution of spectral tilts of murmurs and clear vowels along with their differences (murmur − vowel). It may be seen that the distributions for murmur and adjoining vowel show similar skewed nature with the mode for murmur shifted slightly to the right indicating expected increased negative tilt in general. However the shift of 0.0003 is small, only about 10% of the modal values. It is expected that the spectral structure of a murmured vowel would be influenced by the adjoining vowel. If so the slope of the tilts over different pairs would show good correlation. Unexpectedly the correlation for all data when pooled together is very low only 0.38. It seemed necessary to look into the issue more closely. It is found that the pairs could be clustered to improve correlation. In fact these 246 pairs may be clustered using the differences in the slopes for the pair.

The clustering attempt shows that the pairs of segments with difference of slope ≥ 0.00088 fall into one class (say A containing 116 pairs) and the rest in the other group (say B containing 130 pairs) (Fig. 1.35). Figure 1.35 shows how the pairs of murmured vowels and the adjoining clear vowels tend to group into two separate regions. Though there is some small overlap they are linearly aligned one above the other. Within each group the tilts of murmurs are seen to be strongly correlated in the clusters (0.69 for class A and 0.74 for class B) with those of the adjoining vowels. So far we have been treating all aspirated vowels as murmured vowels. The aforesaid classification reveals an interesting grouping. On listening to these signals by the author himself the two groups of signal produced distinct perceptual difference of the general quality of the aspirated vowels. 81% of these [h] in class A are heard as whispered vowels rest are murmured vowels. Similarly 74% of the [h] in class B are heard as murmured vowels rest are whispered vowels.

Figures 1.36 and 1.37 shows the offset values in percentage namely, 100*(Fc − Fa)/*(Fc + Fa) where Fc, Fa are respectively the given formant frequencies for the aspirated vowel and the adjoining clear vowel. The total number of pairs of

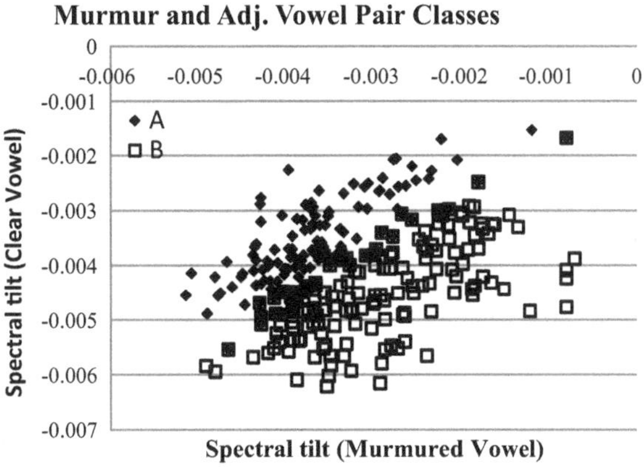

Fig. 1.35 The scatter diagram of spectral tilt of murmur and adjoining vowel pairs

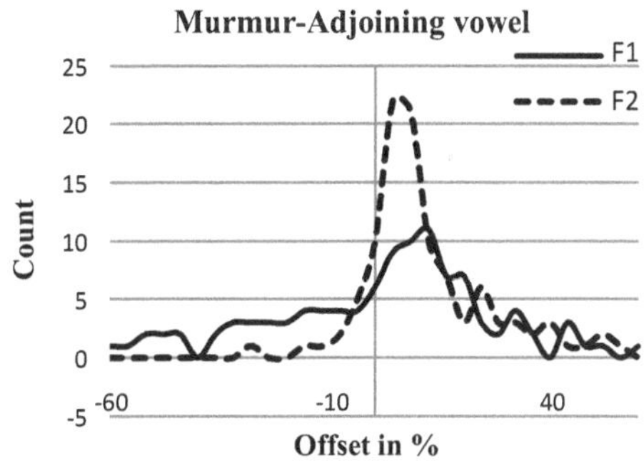

Fig. 1.36 Distribution of offset of murmured over clear vowels

murmured and adjoining clear vowel used in Fig. 1.34 is 105. It may be seen that in general both F_1 and F_2 are higher for murmured vowels. In fact F_2 is higher for 81% of the samples while the corresponding figure for F_1 is 58%. For Fig. 1.35 the number of pairs for the whispered vowels is 135. Here also both for F_1 (82% samples) and F_2 (81% samples) formant frequencies are higher for the whispered vowels. However when we look at the mode of distributions we find that in most cases the shifts are quite small e.g. for murmur these are 10% and 8% respectively for F_1 and F_2. For whispers the corresponding figures are 10% for F_1 and 2% for F_2. Thus aspirated vowels in SCB show consistent raising of lower formant frequencies, albeit by a small amount.

Table 1.6 gives the mean and standard deviations (SD) for the first two formant frequencies of the murmured and whispered vowels. 'n' indicates the number of samples in the category. The SD *values marked bold* indicated reasonably narrow spread. The number of samples in the categories [e, æ, i, and o] are quite small and therefore no comparison would be meaningful. These just indicate that these vowels were also found in the samples available from the corpus. The formants for vowels [ɔ] and [ɐ] are found to be close for the two categories murmurs and whispers. It may be noted that all seven vowels of SCB have the corresponding ones in both

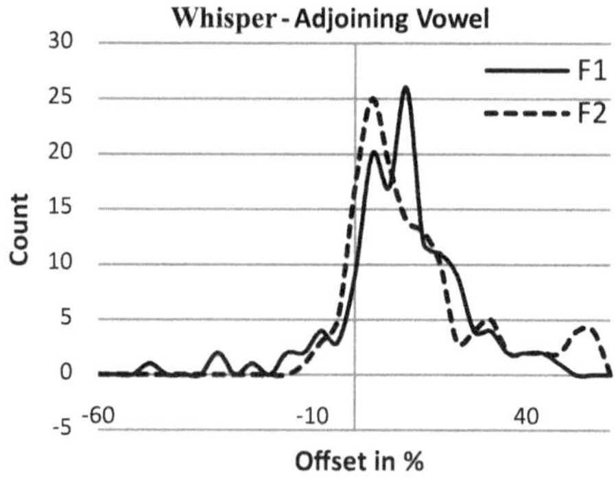

Fig. 1.37 Distribution of offset of whispered over clear vowels

Table 1.6 Mean and SD of different categories of aspirated vowels in SCB

Vowel category	Statistics	Murmured vowels			Whispered vowels		
		F_1	F_2	n	F_1	F_2	n
ɔ	Mean	632	1554	31	800	1646	54
	SD	355	615		213	688	
ɐ	Mean	1006	1915	38	1097	1773	68
	SD	211	452		162	380	
e	Mean	312	2280	11	529	2488	3
	SD	97	227				
æ	Mean	811	1815	8	962	1897	3
	SD	359	316				
i	Mean	387	1974	6	402	2238	4
	SD	286	718				
o	Mean	689	1288	10	360	920	3
	SD	172	422				

murmur and whisper categories even with the small sample size of the present study.

The slope of the periodic decay is the new parameter investigated here. Table 1.7 shows the means and standard deviations of the slopes of the decay for all periods in a selected segment. Average number of periods in a segment is 11 for murmurs and 9 for clear vowels. Altogether 26 pairs of segments have been selected from the database using the following criteria:

(1) Each segment of a pair must have at least 40 ms of reasonably steady state.
(2) The murmured segment should be clear enough for unambiguous periodic segmentation.

Table 1.7 Comparison of slope of periodic decay in murmured and clear vowels

Heard as	Murmured vowel			Adjoining clear vowel			Δ	Heard as
	Mean	SD	n	Mean	SD	n		
ɔ	−37.5	19.3	9	−60.2	14.3	7	23.2	ɔ
o	−30.6	36.9	14	−125.3	42	14	60.7	ɔ
o	−11.5	20.2	7	−20.4	10.6	7	27.9	ɔ
ɔ	−15.3	20.3	5	−80	6.9	7	67.9	ɔ
ɔ	−22	18.2	11	−58.5	26.1	6	45.3	ɔ
ɔ	−23.7	21.2	6	−66.7	29.4	5	47.6	ɔ
ɔ	−38.5	23.6	8	−62.1	26	6	23.5	ɔ
ɐ	−55	23.7	14	−98.6	37.8	12	28.4	ɐ
ɐ	−36.6	13.9	7	−86.4	43.6	11	40.5	ɐ
ɐ	−74.9	51	11	−86.4	43.6	13	7.1	ɐ
ɐ	−36.9	26.8	8	−87.6	25.7	7	40.7	ɐ
ɐ	−34.3	38.9	15	−118.4	71.6	11	55.1	ɐ
ɐ	−11.3	22.4	14	−62.3	31.9	8	69.3	ɐ
ɐ	−14.3	25.3	10	−30.2	15.7	6	35.7	ɐ
ɐ	−30.3	35.7	16	−102	42.8	8	54.2	ɐ
e	−17.5	10.1	7	−104	8.2	4	71.2	e
e	−75.4	30	22	−54.8	17.7	11	−15.8	e
e	−49.9	25.2	11	−12	7.3	10	−61.2	e
e	−13	19.7	11	−131	48.1	8	81.9	e
e	−42.3	20.7	9	−66.2	38.7	7	22.0	e
æ	−51	31.1	11	−81.4	52.7	17	23.0	æ
æ	−27.6	18.7	9	−66.1	32	6	41.1	æ
o	−29.5	22.7	11	−98.5	27.7	8	53.9	o
o	−32.3	20.2	14	−135.5	43.9	11	61.5	o
i	−2.5	51.2	16	−19	10.6	12	76.7	i
i	−23.4	21.5	10	−18.8	13.2	7	−10.9	i

The first and the last column in the table represent the category of the segment as heard by the author; 'n' represents the number of periods in the corresponding segment. 'Δ' is the normalized excess of decay expressed in percentage for clear vowel over the murmured and calculated as $\Delta = 100*(\text{mean clear} - \text{mean murmur})/(\text{mean clear} + \text{mean murmur})$.

It may be seen that the value of Δ is positive in most cases indicating generally that the periodic decay is less for murmurs. Only 3 pairs out of 26 show contrary result. The average value of Δ is 46% indicating that the decrease of the decay rate for murmurs is quite significant. This probably means that the supply of energy from open glottis overrides the possible increase of the damping due to the inclusion of the sub-glottal structure into the resonating system. SD represents the spread of the decay slope around the mean value. A comparison of the SDs with respect to the means for each pair reveals that the spread is considerably high for murmured vowels. On an average it is almost four times larger. This increase in the spread is however expected as in case of the murmurs the vocal folds are said to be more lax causing irregularity in the system to increase.

1.7 Conclusions

Altogether 257 samples of [h] from spoken words were examined. These are taken from 37 words spoken by 7 male and 7 female informants. These may be considered as fairly good samples for having a fare idea of the nature of [h] sound in SCB. Of these 257 samples only 7 are found to be sibilants. The rest appear to consist of 105 murmured and 135 whispered vowels. Contrary to the traditional representation of the sound produced by the grapheme [হ] the study firmly reveals that this grapheme in Bangla is associated with aspirated vowels the vowel quality of which is normally that of the preceding clear vowel. The study firmly reveals the existence of murmured vowels in Bangla corresponding to each of the seven clear vowels.

The characteristic features may be summarized as the following.

- Aspiration noise N is seen to be clearly higher for [ɦ] segments and is a good indicator of murmurs. Also when N is positive, i.e. the spectral energy above 2.8 kHz is more than that below, the vowels are aspirated.
- The lower value of the average power of a murmur/whisper with respect to that of the adjoining vowel is a robust cue for aspirated vowels.
- The differences between the amplitudes of H1 and H2 do not show any discrimination between clear vowels and aspirated vowels.
- The differences in f_0 (adjoining vowel − murmur) is another robust cue for discrimination.
- Though a general trend of increase in the first two formant frequencies are strongly evident the amount is quite small.
- Spectral tilt in the average spectra of the whole segment seems to be a good cue for differentiating aspirated vowels from clear vowels.
- The difference between the spectral tilt of the aspirated vowel and the adjoining clear vowels may be used to differentiate whispered from murmured vowels.
- Strong influences of the adjoining vowel on the murmured as well as whispered vowels are indicated by strong correlation of the spectral tilt.
- Periodic decay, the new parameter proposed is a robust cue for distinguishing murmured from clean vowels. This is significantly faster for murmurs.

1.8 Diphthongs and Semi-Vowels

The other major phonemes in the quasi periodic group of speech sounds are the semi-vowels and the diphthongs. Semi-vowels form a subclass of approximants (Crystal 2003). Semi-vowels, by definition, contrast with vowels by being non-syllabic. In addition, they are usually shorter than vowels. Nevertheless, semi-vowels may be phonemically equivalent with vowels in some situations. Diphthongs are not just a combination of two vowels. Most importantly, diphthongs are fully contained in the syllable nucleus (Schane 1995). While a semi-vowel or glide is restricted to the syllable boundaries (either the onset or the coda), this often manifests itself phonetically by a greater degree of constriction (Padgett 2007). These definitions are linguistic in nature and therefore selection of words is made by a linguist. In this process the linguist uses his own concept of the representative pronunciation of the word. Unfortunately a linguist is a learned person loaded with intricate linguistic knowledge. Such a person's concept of correct utterance could be quite different from that of a common native speaker. In this sense a collection of corpus from speech of native listeners may be different at times when we look only for these conceptual representative words. The concept of syllables is word specific, and words are semantic units. But in continuous speech word boundaries are not clearly marked always. For example let us take the simple spoken query in Bangla, 'ɐmɐrmisţik^{h}ɐben'. If we do not have the supra-segmental information, this utterance has at least two meanings. One is 'would you eat my sweetmeat'. The other is 'would you eat mango and sweetmeat'. In the first case the syllable division would be 'ɐ/mɐr/mis/ţi/k^{h}ɐ/ben'. In the second case it would be 'ɐm/ɐr/mis/ţi/k^{h}ɐ/ben'. Author did not have access to any study in Bangla to indicate whether in real speech syllable definition follows the same rules as in the case of isolated words. Within these constraints the choice of data base for the study on semi-vowels and diphthongs are done. We, therefore, selected all words in the CDAC corpus which is likely to contain one of these phones in addition to the wordlist supplied by the linguists.

The psychoacoustic feature that separates these sounds from the vowels is that the changing spectral patterns reflected in formant movements are innately cognitive. The point of the dynamic movement being cognitive needs a little elaboration. Figure 1.38 shows some nonsense words. Let us concentrate on the last syllables in each of them. The dashed double arrowhead is used to represent the vowel as defined normally. However, in reality it is not so simple. The solid double arrowhead represents the dynamic movement in the resonance structures. These are heard as the corresponding consonants. So they are cognitive but not as a vowel. Only the part of the signal which has a relatively steady formant structure is cognitively a vowel.

In the case of diphthongs this movement indicates the weaker vowel for which the steady state is absent or very short, while in the case of glides the transition alone is the acoustic as well as the cognitive signature.

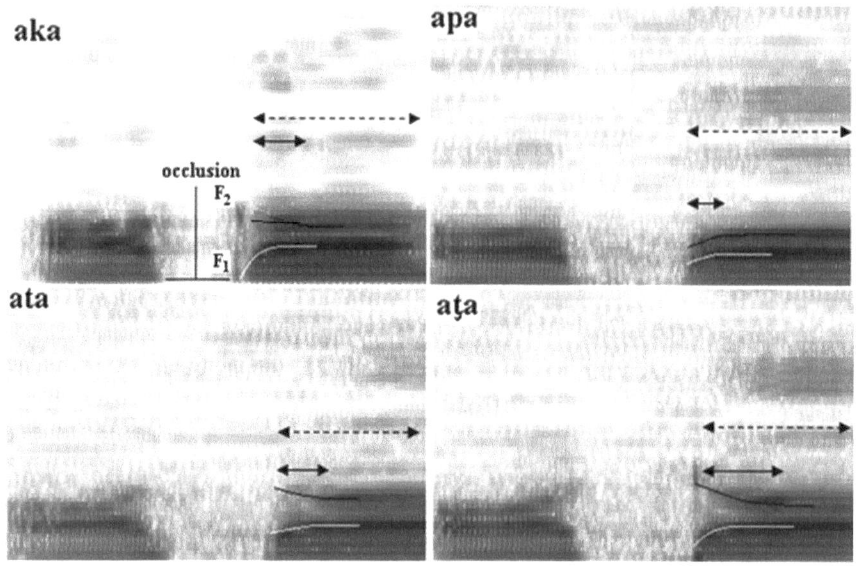

Fig. 1.38 Formant movements in VCV syllables

1.8.1 Diphthongs

The numbers of diphthongs in Bangla cited in literatures vary significantly. Suniti Kumar Chattopadhay (SKC) (Chatterji 1926) cited 25 diphthongs whereas Abdul Hai (AH) (Hai 1989) listed 31 diphthongs, of which 19 are regular and 12 are irregular. Pabitra Sarkar (PS) (Sarkar 1985) has given the number as only 17. Table 1.8 shows the unified list of diphthongs. In this table the phonetic representation follows these authors' descriptions.

Figure 1.39 shows the spectrographic view of a Bangla word containing a diphthong. In this figure, D1 denotes the duration of steady state of 1st vowel [o], D3 denotes the duration of the transitory portion between two vowels [ou] and D2 denotes the duration of steady state of 2nd vowel [u]. Formants automatically detected by *Wavesurfer* are indicated by curved lines of different colors.

1.8.2 Acoustic Signatures

As already mentioned the semi-vowels and diphthongs reveal characteristic transitions of acoustical parameters contained in the quasi-periodic acoustic waves. Segments corresponding to each word containing semi-vowels and diphthongs had been extracted manually from the corpus [CDAC, 49]. The durations of the steady states of initial and final vowels, along with that of transitions are also manually extracted from the spectrogram. Figure 1.40 presents 3D spectrographs of 25

Table 1.8 Unified list of diphthongs with example words

	SKC	AH	PS	Unified list	Example word	IPA
1	*ɔa*	*ɔa*	–	*ɔa*	নয়া	nɔa
2	*ɔe*	*ɔj*[1]	*ɔj*[1]	*ɔe*	হয়	hɔe
3	*ɔo*	*ɔo*	*ɔo*	*ɔo*	বও	bɔo
4	–	*æa*	–	*æa*	ন্যায়াধীশ	næad^{h}iʃ
5	*æe*	*æj*[1]	*æj*[1]	*æe*	নেয়	næe
6	*æo*	*æo*	*æo*	*æo*	শ্যাওলা	ʃæola
7	*ai*	*ai*	*ai*	*ai*	ভাই	bhai
8	*ae*	*aj*[1]	*aj*[1]	*ae*	খায়	khae
9	*ao*	*ao*	*ao*	*ao*	খাও	khao
10	*au*	*au*	*au*	*au*	লাউ	lau
11	*ea*	*ea*	–	*ea*	খেয়া	k^{h}ea
12	*ei*	*ei*	*ei*	*ei*	সেই	sei
13	–	*ejo*	–	*ejo*	চেয়ো	ʧeo
14	*eo*	*eo*	–	*eo*	শেও	ʃeo
15	*eu*	*eu*	*eu*	*eu*	ঢেউ	d^{h}eu
16	*ia*	*ia*	–	*ia*	প্রিয়া	pria
17	*ie*	*ie*	–	*ie*	গিয়ে	gie
18	–	*ii*	*ii*	*ii*	দিই	dii
19	*io*	*io*	–	*io*	নিও	nio
20	*iu*	*iu*	*iu*	*iu*	বিউলি	biuli
21	*oa*	*oa*	–	*oa*	নোয়া	noa
22	–	*oe*	–	*oe*	সয়ে	sɔe
23	–	*oi*	*oi*	*oi*	বই	boi
24	*oe*	*oj*[1]	*oj*[1]	*oe*	ধোয়	d^{h}oe
25	–	*oo*	*oo*	*oo*	শোও	ʃoo
26	*ou*	*ou*	*ou*	*ou*	নৌকা	nouka
27	*ua*	*ua*	–	*ua*	জুয়া	ʤua
28	*ue*	*ue*	–	*ue*	শুয়ে	ʃue
29	*ui*	*ui*	*ui*	*ui*	দুই	dui
30	*uo*	*uo*	–	*uo*	থুয়ো	t^{h}uo
31	–	*uu*	–	*uu*	কুউ	kuu

[1]It appears that the corresponding authors took some liberty in presenting a diphthong as combination of one vowel and a glide instead of generally accepted presentation with two vowels

different diphthongs. Each one of them reveals the strong characteristic transitions. As we could not find any word containing [æi] and [iɔ] in SCB so data for these two diphthongs could not be included. Further it was noticed that some Bangla words

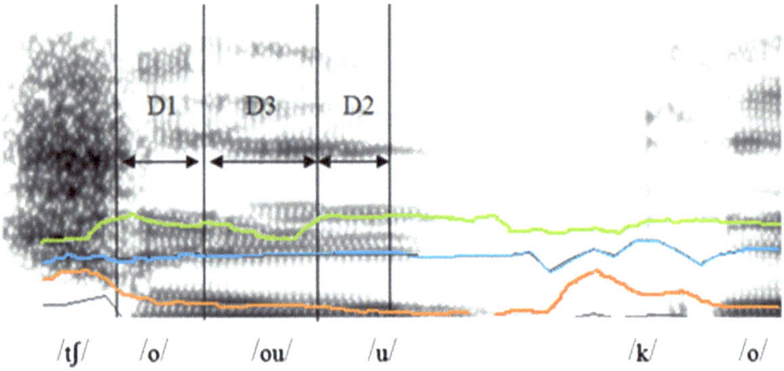

Fig. 1.39 Spectrographic view of Bangla word (/tʃouko/)

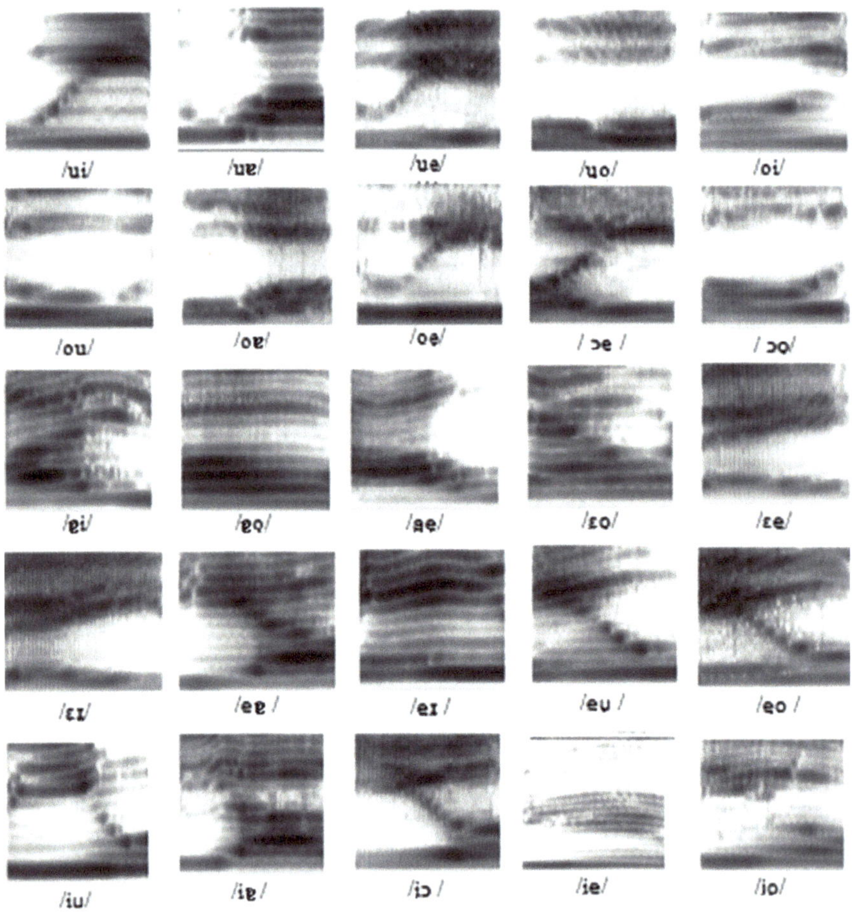

Fig. 1.40 Spectrogram of Bangla diphthongs

contain vowel pairs which though not listed in Table 1.8 need to be examined to see whether they have the acoustic characteristic of a diphthong. These include pairs [*ɐe*], [*œa*], [*ɔɐ*], [*ejo*], [*ii*], [*oj*], and [*uu*]. The spectrograms of the diphthongs [uu], [oo] and [ii] as well as of [ɔɐ], [ɐu], [æɐ], will be discussed later along with other glides.

From the individual spectrograms in Fig. 1.40 one can apparently see that the first vowel is weak for ui, uɐ, ue, oɐ, oe, ɔe, ɐi, ei, iɐ, io and the second vowel is weak for oi, ou, ɔo, ɐe, æi, eɐ, eu. However these are first impression only. The weaker vowel would be properly identified from the acoustic parameter later on. Lehiste and Peterson (1961) suggested representation of diphthongs by superscripting the weak vowel.

1.8.3 Results and Discussions

Table 1.9 presents the average values and the standard deviations of the durations of the steady states of the related two vowels and those of the in-between V-V transition for all the combinations under study. The standard deviation show reasonable variation expected from a natural source. The data is not large enough for a formal expression on the statistics. The figures in parenthesis give the percentage of duration of the item to that of the whole phone. The entries are organised in increasing order of the duration of the first target vowels.

One interesting point that revealed from the data is that in Bangla diphthongs except for [oe], [iɔ] and [æo] both the vowels have significantly long duration to be heard as a clear vowel if we consider 40 millisecond duration to be minimum requirement of cognition of a vowel. They also have long duration for transition. However traditionally these are referred as single syllabic nucleus.

However if we see the relative duration then percentage of the transitional duration to the total duration of a nucleus the average for Bangla is 42%. This is quite close to those reported for American English (Lehiste and Peterson 1961). For diphthongs [au], [aɪ] and [ɔɪ] in American English the transition duration is about 41% of the total duration while the durations of other vowels are around 36% and 23% respectively for the first and second target vowels. Very high proportional duration has been noticed for Romanian diphthongs [ja], [ea], [wa] and [oa]. The transition duration is reported to be rather large (70–80% of the total duration) (Chitoran 2002).

Some of the aforesaid diphthongs in cited references like [uo], and [ɔo] respectively in words like থুয়ো (t^huo), and নয়া (nɔa) have almost equal percentage of durations for the three constituent segments namely, the two vowel and the intermediate transition. The durations of the segments are all more than 65 ms each and, therefore much above the cognitive threshold. Yet they are reported to be a single nucleus and considered as a diphthong. It is surprising that [t^huo] and [nɔa] cannot be divided each into two syllables namely, [t^hu-wo] and [nɔ-ja], instead has to be represented as [t^hu^o] and [nɔa].

We can use the durational values to prune the aforesaid list given in Table 1.8. If we assume that a pair of vowels can be considered as a diphthong only when at

Table 1.9 Average duration and standard deviation of all the diphthongs

Diphthongs (traditional view)	No of samples	Duration in milliseconds					
		First vowel (steady state)		Transition		Second vowel (steady state)	
		Average	SD	Average	SD	Average	SD
[oe]	15	37.67 (19.01)	16.95	79.93 (40.33)	24.9	80.6 (40.67)	28.88
[iɔ]	11	37.8 (24.71)	11.75	66.4 (43.4)	14.64	48.8(31.9)	10.4
[io]	25	41.52 (26.19)	14.79	73.8 (46.56)	16.62	43.2 (27.25)	18.81
[oi]	19	42.95 (25.92)	17.25	75 (45.27)	22.02	47.74 (28.81)	20.15
[ui]	11	43.64 (24.37)	13.80	76.55 (42.74)	15.91	58.91 (32.89)	24.24
[ai]	25	44 (25.58)	18.72	79.72 (46.35)	18.3	48.28 (28.07)	16.09
[ei]	10	44 (29.2)	15.45	61.6 (40.88)	21.34	45.1 (29.93)	22.17
[ue]	12	44.33 (21.32)	18.13	78.17 (37.6)	21.37	85.42 (41.08)	28.44
[iu]	10	46.3 (26.49)	24.93	64.5 (36.9)	11.41	64 (36.61)	24.35
[ɔe]	8	50.5 (23.6)	17.6	92.38 (43.17)	14.32	71.13 (33.24)	17.21
[æo]	7	51 (28.24)	12.83	91.14 (50.47)	34.68	38.43 (21.28)	10.03
[i.e.]	28	51.5 (31.36)	23.89	70.21 (42.72)	18.52	42.61 (25.92)	20.34
[eu]	17	52.89 (25.92)	18.2	78.06 (38.25)	18.06	73.12 (35.83)	21.6
[eo]	15	53.8 (24.92)	18.45	96.47 (44.69)	23.53	65.6 (30.39)	17.33
[ia]	9	55.67 (27.51)	11.92	91.56 (45.25)	18.17	55.11 (27.24)	15.74
[ua]	12	56.17 (28.27)	18.15	78 (39.26)	21.53	64.5 (32.47)	26.72
[ea]	14	61.5 (28.93)	21.18	98.93 (46.54)	23.26	52.14 (24.53)	48.88
[æi]	10	61.7 (31.71)	18.2	72.8 (37.41)	24.07	60.1 (30.88)	15.59
[æe]	4	64.75 (27.38)	2.06	107.25 (45.35)	17.46	64.5 (27.27)	5.26
[oa]	4	73 (31.43)	12.11	95 (40.9)	22.46	64.25 (27.66)	14.2
[ɔo]	5	76 (33.45)	33.42	83.2 (36.62)	21.8	68 (29.93)	48.41
[uo]	7	70.86 (31.86)	26.63	84.29 (37.89)	13.03	67.29 (30.25)	22.09
[ao]	13	55.77 (31.98)	24.13	72.15 (41.38)	19.6	46.46 (26.64)	19.33
[au]	15	76.73 (32.64)	23.84	77.2 (32.84)	31.35	81.13 (34.51)	27.02
[ou]	11	66.64 (35.43)	27.62	74.09 (39.39)	21.87	47.36 (25.18)	23.12
Total	317	(24.96)		(42.09)		(30.37)	

Table 1.10 Average duration and standard deviation of selected diphthongs

Diphthongs	Duration in milliseconds		
	First vowel	Transition	Second vowel
	(Steady state)		(Steady state)
	Average	Average	Average
[ei]	44 (29.2)	61.6 (40.88)	45.1 (29.93)
[iu]	46.3 (26.49)	64.5 (36.9)	64 (36.61)
[i.e.]	51.5 (31.36)	70.21 (42.72)	42.61 (25.92)
[ao]	55.77 (31.98)	72.15 (41.38)	46.46 (26.64)
[æi]	61.7 (31.71)	72.8 (37.41)	60.1 (30.88)
[io]	41.52 (26.19)	73.8 (46.56)	43.2 (27.25)
[ou]	66.64 (35.43)	74.09 (39.39)	47.36 (25.18)
[oi]	42.95 (25.92)	75 (45.27)	47.74 (28.81)
[ui]	43.64 (24.37)	76.55 (42.74)	58.91 (32.89)
[au]	76.73 (32.64)	77.2 (32.84)	81.13 (34.51)
[ua]	56.17 (28.27)	78 (39.26)	64.5 (32.47)
[eu]	52.89 (25.92)	78.06 (38.25)	73.12 (35.83)
[ue]	44.33 (21.32)	78.17 (37.6)	85.42 (41.08)
[ai]	44 (25.58)	79.72 (46.35)	48.28 (28.07)

least one of them is weak and below the threshold of cognition and the cognitive threshold of duration to be taken to be at least 40 ms. then only 22 pairs from the aforesaid list qualify. Of these 8 have transitional duration higher than 80 ms which is much more than required for cognition of it as a semi-vowel. Thus the acoustic data suggest that of all the vowel pairs given in Table 1.9 only 14 qualifies acoustically the conditions required for a diphthong. These are presented in Table 1.10 (figures within parenthesis represent the standard deviations). Of these 10 (in bold script) matches with the list cited by PS (see Table 1.8).

Furthermore a large number of the V-V combinations particularly, [ɔe], [ia], [ea], [æe], [oa], [eo], [æo], [uo] and [ɔo] in the list have very large transition duration indicative of a glide. Listening experiments need to be done to ascertain their character.

1.9 Semi-Vowels

A glide or a semi-vowel is a vocalic syllable nucleus consisting of a single target vowel. During the production of a glide the tongue forms a constriction just wide enough not to cause turbulent airflow. The associated tongue movement is comparatively slow and causes formant transitions to or from the target vowel a cognizable event by itself. The duration of movement is comparable to that of the target. Cognitively the sound is heard as a gliding sound.

Traditionally it is believed that Bangla has three semi-vowels or glides namely /j/, /y/ and /w/ (Hai 1989). These pertain to Bengali words and have characteristic signatures of movement in the spectral domain. However if we take into account continuous speech such glides may also occur at the word juncture when the preceding word ends in a vowel and the succeeding word also has a vowel in the beginning. In such cases when the two vowels are same a pitch glide is very common (Datta et al. 1989). A question may be raised at this juncture as to why should one bother about a speech sound which is not a part of the word and therefore not a phoneme. However as this is a work on acoustics which relates to spoken language not to textual language one has to be aware of all speech events which have cognitive connotation while knowing well what one is looking at. As has already been pointed out earlier, that words in speech do not consistently leave special markers for a word-ending. In fact the recent trends in spoken language processing is not to be obsessed with the abstract unit called word but to give more emphasis on what is usually referred to as prosodic words. These are usually small phrases having consistent prosodic markers at the signal level. In any case any additional knowledge cannot be harmful.

One may note here that for studying the acoustic signatures one need to use semi-vowel in all possible context available in the spoken database. We note that in combination of the type vowel-semi-vowel-vowel normally occurring in lexicon the purported semi-vowel could be a syllabic boundary. It could also be part of a diphthong. The present purpose is to study the acoustic signature of transitory movements of acoustic parameters of certain phones without in any way interfering with the cognitive evaluation of erudite linguists. Such studies have earlier been reported in other languages which, no doubt, augmented our knowledge of acoustic phonetics. A large number of spoken words were carefully examined to select 202 samples where glides are heard clearly. As this data comes from 10 speakers of both sexes and as the data was not large enough to have a speaker specific analysis the data were pooled together for a general analysis.

1.9.1 Acoustic Signatures

The general acoustic signature of a glide is a long transition either preceded or followed by cognizable vowel. The following acoustic signatures were reported (Ganguly et al. 1999) for glides in SCB:

1. Existence of virtual target ([kɐjɐ])
2. Single target vowel ([koetʃi], [keukeʈɐ]
3. Two targets connected by a transition ([gɐonɐ], [sɔodɐ]
4. An additional short target ([pɐj]
5. Pitch glide ([beje], [peje].

Figure 1.41 presents the examples of acoustic structures of glides in different vowel contexts. In the figure glides are represented by 'x' instead of their traditional symbols. Virtual targets have been seen for combinations [ɐxɐ], [ɐxo], [ɐxu], [ɔxɐ], [ɔxɔ], [oxu], [oxo], [exɐ] and [ɔxɐ]. The actual symbols for this will be presented later in Table 1.11. Of these the last two are when F_2 goes low and the rest are when F_2 goes high. The targets are said to be virtual when they are not heard as vowels though in the spectrograms they present very distinctive features of vowels. The other interesting point to note that in general virtual targets are created when the two end vowels have very close frequency for the second formant. It is quite possible that these are generated to provide additional support to the cognitive mechanism for the new phonetic event. Figures for combinations [ɐx], [æx], [ɔx], [ox] and [xɐ] are examples of glides associated with single target vowel similarly [æxɐ], [oxo], [cxo] and [ɔxɐ] may be considered as glides between two target vowels.

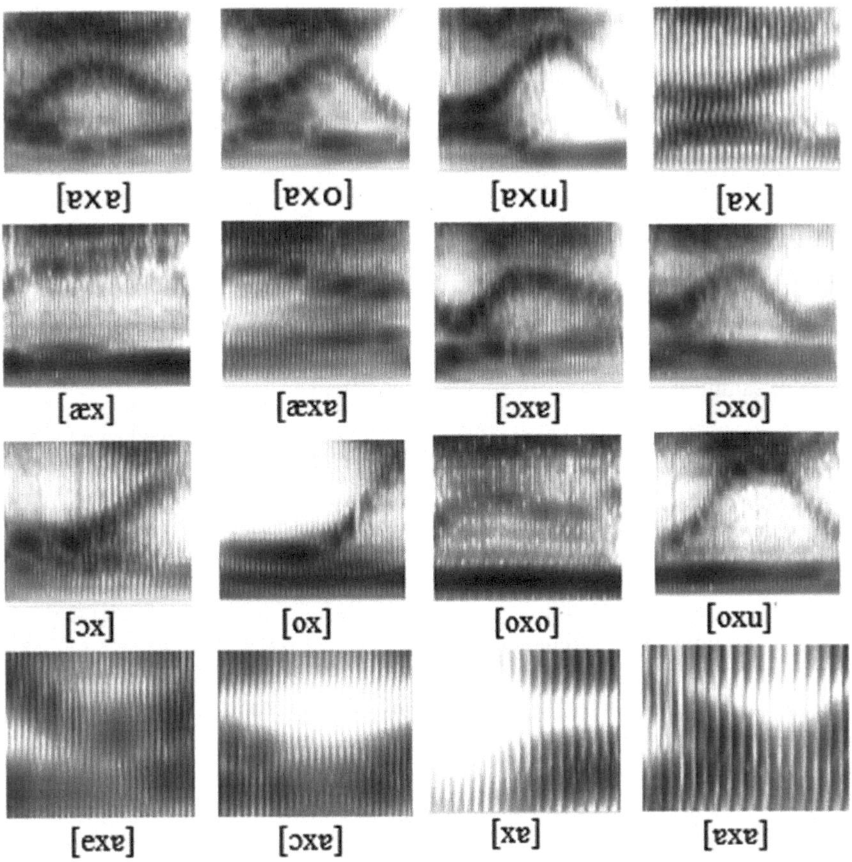

Fig. 1.41 Spectrogram of Bangla semi-vowels in different vowel contexts

Table 1.11 Tentative categorization of SCB semi-vowels

Category	Symbol	SCB combinations in words
Palatal	[j]	[aja], [ajo], [aju], [æja], [ɔja], [ɔjo]
Labio-palatal	[ɥ]	[oɥu], [oɥo]
Velar	[ɰ]	[eɰa], [aɰa]
Labio-velar	[w]	[ɔwa], [wa]

It is also desirable to attempt at fixing the symbols for the glides with reference to the spectral signatures presented above. It may be noted that spectral signatures of all examined words are not presented above. Only one word from each group is selected as an example. If we consider that high F_2 is a signature of palatalization as against low F_2 for a signature for velar constriction at the virtual targets the glides represented in Fig. 1.41 may be represented in Table 1.11. Also if the adjoining vowels are rounded ones it is logical to select the symbol for labial counterpart. Using this approach one finds the existence of all the four semi-vowels in SCB. However this categorization is only tentative and there is a need to conduct listening experiments involving phoneticians to ascertain the veracity of such categorization.

The study reveals some gliding sounds in case of some of the Vowel-Vowel combinations consisting of same vowel like [ii], [oo], [uu] and [ee] respectively in words [dii] (meaning *I give*), [ʃoo] meaning *lie down*), [kuu] (meaning *call of bird cuckoo*), [tʃee] (meaning *after seeing*) where gliding sounds are clearly heard. In such cases only the duration of pitch glide is noted, which is shown in Fig. 1.42. The pitch contour is shown by black line. The continuously changing value of the pitch causes the gliding sensation though there is no significant change in spectra during this glide. The duration was noted to be long enough to cause the glide sensation. This phenomenon was found to occur also at word juncture e.g., [ma amar] (meaning *mother mine*) (Ganguly et al. 1999).

1.9.2 Duration

Tables 1.12 and 1.13 present the durations of different acoustic segments of the traditionally reported single vowel and semi-vowel combination, and vowel-semi-vowel-vowel combination respectively. It may be seen from the table that the duration of transitions representing glides are generally longer than the target vowels. In fact in most of the cases they are large enough to be a cognitively relevant entity. It may be noticed that [w] has a comparatively shorter transition than [j] but still it is large enough to be cognitively relevant and is larger than the vowel duration.

In Table 1.13 the virtual targets are estimated on the basis of average F_1 and F_2 values. In general the transition is much larger than the targets. However, for [ɔjɔ] and [ojo] the first targets show very short durations. Like in the instances in V-j

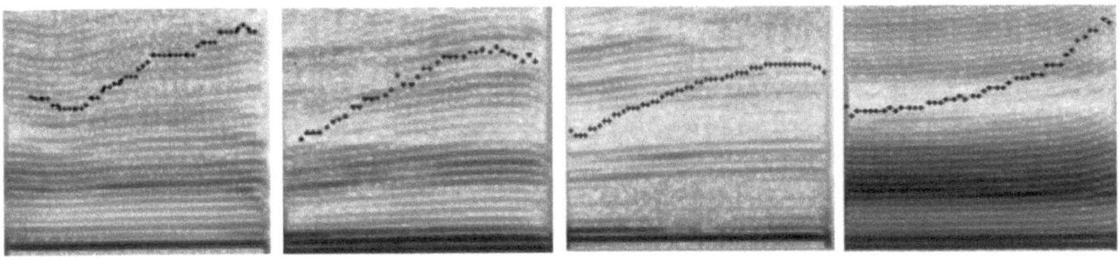

Fig. 1.42 Examples of pitch glide in between two vowels of same category

combinations, in all instances of V-j-V combinations the durations of transitions seem to define glides without doubt.

1.10 Discussions

Standard colloquial Bengali commonly known as Bangla is now used in formal and cultured mass communication throughout West Bengal. The study of phonetics in it has a long history, the early ones being purely subjective. These represented phonetics of SCB as imitated by erudite scholars of whatever nativity they may be. It is natural that these would be very subjective and useful for scholarly studies. However in the present era where phonetics is the most important part of technology development in voiced communication it is imperative to capture acoustic phonetics of speech sounds, which represents the native speaking as closely as possible. Not only that these have to be represented in terms of the physics of sound

Table 1.12 Average duration and standard deviation vowel and adjoining semi-vowel

V-j combination	No. of samples	Duration (in ms)			
		First vowel steady state		Transition of glide	
		Average	Standard deviation	Average	Standard deviation
[æj]	10	116.9	23.34	111.9	12.48
[aj]	18	84.22	27.72	130.28	28.03
[ɔj]	19	66.26	20.59	133.84	27.63
[oj]	10	63.2	16.42	133.5	26.15
		Duration(in ms)			
w-V combination		Transition with semi-vowel (w)		Second vowel steady state	
		Average	Standard deviation	Average	Standard deviation
[wa]	10	88	16.68	38.80	10.08

Table 1.13 Average duration and SD of all the vowel-semi-vowel-vowel combinations

V-j-V	No of samples	Virtual target	Duration (in ms)					
			First vowel		Transition		Second vowel	
			Average steady state	Standard deviation	Average	Standard deviation	Average steady state	Standard deviation
[æja]	4	e	72.25	27.02	106.25	26.51	53.5	6.76
[aja]	19	æ	51.79	13.21	165.37	40.82	57.74	17.44
[ajo]	17	e	58.88	17.65	155.41	31.03	48	19.69
[aju]	15	i	47.53	11.19	185.73	28.5	61	19.95
[ɔja]	15	æ	49.4	11.25	174.67	38.05	67.33	23.89
[ɔjɔ]	6	e	35.83	5.12	148	25.47	48.17	12.14
[ɔjo]	20	e	56.2	19.13	143.95	27.12	59	17.86
[ojo]	4	i	33.5	12.4	149.75	13.07	47	18.89
[oju]	5	i	46.8	5.89	169.4	33.69	77.2	22.95
V-w-V								
[ɔwa]	10	o	63.4	15.15	86.4	17.37	59.7	19.3
[awa]	10	o	39.92	14.7	88.81	40.8	29.6	19.92
[ewa]	10	o	11.06	103.36	16.52	43.91	14.65	11.06

so that measures may be developed for each of them for the use of technology. This development is not an ornamental ideational exercise. It has firm need for the society where literacy is not much above 70% and the functional literacy rate is even much lower. The estimate of functional literacy is no common task and usually such reports are not above serious controversies. In general one can assume that not more than 40–50% people in India can read a general informative passage and glean the needed knowledge content out of it. Unfortunately the rest i.e. the so handicapped people consists of the population engaged in production of mass consumables. They need to be brought under the knowledge resources needed for upgrading their standard of life as well as bringing efficiency in production. The potential of speech technology lies here. It is now possible to bring knowledge to these people of India. The instruments for oral communication with computers through speech mode are ready. This consists of two separate engines: (a) a computer-man communication engine commonly known as text to speech synthesis engine (TTS) and (b) a man-machine communication engine called automatic speech recognition (ASR) engine. Fortunately indigenously developed TTS engines are now available. One may refer to ESNOLA based speech synthesizer (Choudhury 2006) and ESOLA based engine (Das Mandal 2007). A new engine called Manner Based Lexically Driven (MBLD) engine is developed for isolated word recognition (Das Mandal 2007). But these require language specific knowledge modules. This knowledge has two components, semiotics and acoustics. Again the semiotic knowledge, consisting of syntax, grammar, pragmatics, is mostly available for Bangla. But these are mainly based on textual language. To what extent these conform to spoken dialect has not been investigated. One would expect considerable deviations and therefore needs necessary modifications to fit the need of technology. Fortunately the acoustic knowledge is being overhauled for consumption by the technology. The subjective abstract knowledge of phones and their contextual interactions need to be precisely and objectively specified. It has to be field dependent not simply an ideational exercise.

This means a new paradigm of research using scientific tools of measurements and the technology of statistics to choose representative from a collection of large number of data has to be ushered in. Obviously one expects some changes in the final representation. In India this has began late last century. The present work though primarily a recent study with collection of contemporary speech data also includes the results of earlier work conducted in the country. An attempt has been made above to provide purely objective data base for use in the technology. This, we feel, may also be useful for subjective ideational review of the existing knowledge. Attention may be drawn to two new findings in the present work. These are the existence of aspirated vowels to replace the present linguistic gross notion of sound corresponding to the grapheme [h] and a new set of glides including the inclusion of pitch glides. One may also like to draw attention to the observed new set of diphthongs which has a firm acoustic footing and may help in developing the TTS and ASR engines.

A notable addition is the use of a very recent development which allows one to use objectively collected data, namely the first two formant frequencies and the fundamental frequency, to accurately estimate the perceptual subjective categorization of vowels. Interestingly this produced a more consistent and compact distribution of vowels in parametric plane. This has strong potential in automatic speech recognition apart from being an example of interaction between hard and soft sciences.

References

Andruski JE, Ratliff M (2000) Phonation types in production of phonological tone the case of Green Mong. J Int Phon Assoc 30:39–62

Berkins S, Stevens KN (1982) Across language study of the perception of nasal vowels. J Acoust Am 73(suppl 1):s 54

Bickley C (1982) Acoustic analysis and perception of breathy vowels. In: Speech communication group working papers. Cambridge Massachusetts Institute of Technology, pp 71–82

Biswas S (2004) Samsad Bangla Dictionary (Samsada Bangala Abhidhana), 7th ed. Calcutta, Sahitya Samsad

Chatterji SK (1926) The origin and development of the Bengali language. Rupa & Co, New Delhi

Chávez-Peón ME (2013) Non-modal phonation in Quiaviní Zapotec: an acoustic investigation. Instituto de Investigaciones Antropológicas, Universidad Nacional Autónoma de México. Retrieved 26 May 2013

Chitoran I (2002) A perception-production study of Romanian diphthongs and glide-vowel sequences. J Int Phon Assoc 32(2):203–221

Choudhury S (2006) Concatenative text to speech synthesis a study of standard colloquial Bengali. PhD Thesis, Indian Statistical Institute

Crystal D (2003) A dictionary of linguistics and phonetics, 5th edn. Wiley-Blackwell, ISBN 0-631-22664-8

Das Mandal SK (2007) Role of shape domain parameters in speech recognition a study on standard colloquial Bangla (SCB). PhD Thesis, Jadavpur University

Datta AK (2014) Aspirated vowels in standard colloquial Bengali: a case study with native informants communicated to JPH, Springer

Datta AK, Ganguly NR, Dutta Majumder D (1978) Some studies on acoustic features of Telugu vowels. Acustica 41:55–64

Datta AK, Ganguly NR, Mukherjee B (1989) Bengali nasal sounds—a spectrographic study. J Acoust Soc India XVII:219–223

Datta AK, Sengupta R, Dey N, Banerjee BM, Nag D (1998) Perception of nasality in Bengali vowels role of harmonics between F0 and F1. In: Proceedings of international conference on computational linguistics, speech and document processing, ISI, Calcutta, 18–20 Feb 1998

Delattre PC, Liberman AM, Cooper FS, Gerstman LJ (1952) An experimental study of the acoustic determinants of vowel colour; observations on one- and two-formant vowels synthesized from spectrographic patterns. Word 8(3):195–210

Djordje K, Das R (1972) Short outline of Bengali phonetics. Statistical Publishing Society, Calcutta

Dutta AK, Sengupta R, Dey N (2003) Jitter, Shimmer and HNR characteristics of singers and non-singers. J ITC Sangeet Res Acad 17

Dutta Majumder D, Datta AK (1966) A scheme for automatic speech coding and recognition. In: ISALS symposium on control computation

Esposito CM, Khan SUD, Hurst A (2007) Breathy nasals and /Nh/ clusters in Bengali, Hindi, and Marathi. UCLA working papers in phonetics, vol 104, pp 82–106

Fant G (1970) Acoustic theory of speech production. Mouton De Gruyter

Fant G, Ishizaka K, Lindquist-Gauffin J, Sundberg J (1972) Subglottal formants, STL-QPSR 1/1972
Fischer-Jorgensen E (1967) Phonetic analysis of breathy (murmured) vowels in Gujarati. Indian Linguist 28:71–139
Fujimara O (1960) Spectra of nasalised vowels. Res Lab Electron Q Prog Rep 58, MIT 214–218
Ganguly NR, Datta AK, Mukherjee B (1988) Acoustic phonetics of non-nasal standard Bengali vowels a spectrographic study. J Electron Telecommun Eng 34(1):50–56
Ganguly NR, Datta AK, Mukherjee B (1999) Acoustic phonetic features of glides and diphthongs in Bengali. J Acoust Soc Ind XXVII:199–202
Gordon M (2001) Phonation types: a cross-linguistic overview. J Phon
Hai AM (1989) Dhani Bigyan and Bangla Dhani Tattwa. Mallick Brothers, Dhaka, Bangladesh
Hanson H (1995) Glottal characteristics of female speakers. PhD dissertation, Harvard University, MA
Hartmut RP (2003) Acoustic correlates of the IPA vowel diagram. In: 15 ICPhS Barcelona, pp 1441–1444
Hawkins S, Stenens KN (1985) Acoustic and perceptual correlates of the non-nasal-nasal distinction for vowels. J Acoust Am 77(4):1560
Hillenbrand J, Houde RA (1996) Acoustic correlates of breathy vocal quality dysphonic voices and continuous speech. J Speech Hear Res 39:311–321
Hillenbrand J, Cleveland RA, Erickson RL (1994) Acoustic correlates of breathy vocal quality. J Speech Hear Res 37:769–778
Hombert JM, Ohala J, Ewan W (1979) Phonetic explanations for the development of tones. Language 55(1):37–58
Hossain SA, Rahman ML, Ahmed F (2005) Acoustic space of Bangla vowels. In: WSEAS 5th international conference on speech and image processing, Greece, Aug 2005, pp 138–142
http://www.cdackolkata.in/html/txttospeeh/corpora/corpora_main/MainB.html
http://www.phonetics.ucla.edu/course/chapter1/vowels.html
http://www.speech.kth.se/wavesurfer
Jones D (1962) An outline of English phonetics. W. Heffer & Sons Ltd., Cambridge
Klatt DH, Klatt C (1990) Analysis, synthesis, and perception of voice quality variations among female and male talkers. J Acoust Soc Am 87(2):820–857
Konno H, Kanemitsu H, Toyama J, Shimbo M (2006) Spectral properties of Japanese whispered vowels referred to pitch. J Acoust Soc Am 120(5):3378
Ladefoged P (1967) Three areas of experimental phonetics. Oxford University Press, Oxford
Ladefoged P, Antananzas-Barroso N (1985) Computer measures of breathy voice quality. UCLA working papers in phonetics, pp 61, 79–86
Laver J (1980) The phonetic description of voice quality. Cambridge university press, Cambridge
Lehiste I, Peterson GE (1961) Transitions, glides, and diphthongs. JASA 33(3):268–277
Masica CP (1991) Indo-Aryan languages. Cambridge University Press, Cambridge
Mistry PJ (1997) Gujarati phonology. In: Kaye AS (ed) Phonologies of Asia and Africa. Winona Lake Eisenbrauns
Ngoc YPT, Badin P (1994) Vocal tract acoustic transfer function measurements further developments and applications. Supplement au Journal de Physique 111, vol 4, May 1994
Padgett J (2007) Glides, vowels, and features. Lingua 118(12):1937–1955
Pandit PB (1954) Indo Aryan sibilants in Gujarati. Indian Linguisties 14
Saksena BR (1971) Evolution of Awadi. Motilal Banarasi Das Publication, New Delhi, pp 74–76
Sarkar P (1985) Bangla Dwishardhani (Bangla Diphthong), Calcutta, 1985–86
Schane S (1995) Diphthongization in particle phonology. In: Goldsmith JA (ed) The handbook of phonological theory, Blackwell handbooks in linguistics. Blackwell, pp 586–608
Stevens K (2000) Acoustic phonetics. Massachusetts MIT Press
Stevens K, Hanson H (1994) Classification of glottal vibration from acoustic measurements. Paper presented at the 8th vocal fold physiology conference, Kurume, Japan, 7–9 Apr 1994
Takeuchi S, Kasuya H, Kido K (1975) On the acoustic correlates of nasality. J Acoust Jpn 31:298–309

Tartter VC (1989) What's in a whisper? J Acoust Soc Am 86(5):1678–1683
Teagre HM, Teagre SM (1990) A phenomenological model for vowel production in the vocal tract. In: Daniloff RG (ed) Speech science recent advances. College Hill, San Diago, pp 73–109
Thongkum T (1988) Phonation types in Mon-Khmer languages. In: Fujimura O (ed) Vocal fold physiology voice production, mechanisms and functions. New York Raven Press, pp 319–334
Tsunoda K, Ohta Y, Soda Y, Niimi S, Hirose H (1997) Laryngeal adjustment in whispering magnetic resonance imaging study. Ann Otol Rhinol Laryngol 106:41–43
Varshney RL (1995) An introductory textbook of linguistics and phonetics, 8th edn. Student Store
Wayland R, Jongman A (2003) Acoustic correlates of breathy and clear vowels the case of Khmer. J Phon 31:181–201

Chapter 2
Consonants

2.1 Introduction

Bangla consonants are broadly classified into plosives, affricates, fricatives, lateral, taps, and trill according to their manners of production. Again in each of these manner-based categories there could be categorization on the basis of the place of articulation. All these traditional categorizations of consonants in SCB are shown in Table 2.1 with example words for consonants in initial, medial and final positions, as are available in literatures (Chatterji 1926; Hai 1989).

The aforesaid categorization was done by expert phoneticians and linguists.

However to suit the need of technology, objective quantifiable acoustic parameters as exists in today's speech sound is imperative. Fortunately, in recent times, with the advent of technology it is possible to objectively and quantifiably determine the manners and places of articulation of a particular consonant uttered a number of times by a number of speakers. Statistics allows us methodology to select the appropriate representative. Such objective categorization is necessary for the development of speech technology like ASR and TTS engines. As has been said earlier in India this type of activity began in early 70s (Datta et al. 1974) in Kolkata with the procurement of the first Sonagraph from Kay Elemetrics. However the work on Bangla started much later.

2.2 Experimental Procedure

Some of the instrumental approaches like signal processing tools including extraction of spectral structures have been in use in India since 70s. The use of spirometers, electroglottograms (EGG), palatographs, etc., is also quite old. A new entrant in this field is the electropalatograph (EPG). This is now in use in India. While the spirometers and electroglottograms are used for detecting the manner

A.K. Datta, *Acoustics of Bangla Speech Sounds*, Signals and Communication Technology, DOI 10.1007/978-981-10-4262-1_2

Table 2.1 Place and manner of articulation of the Bangla consonant

S. No.	Grapheme	IPA symbol	Example Bangla words with phoneme in			Place of articulation	Manner of articulation
			Initial	Medial	Final		
1	ক	/k/	/kobitɐ/- 'Poetry'	/kɐkɐ/ 'Uncle'	/ɔbɐk/ 'Surprise'	Velar	Unaspirated unvoiced stop
2	খ	/kʰ/	/khɔtom/ 'End'	/ɔkhil/ 'Whole'	/ɐkh/ 'Sugar cane'	Velar	Aspirated unvoiced stop
3	গ	/g/	/gɐdhɐ/ 'Donkey'	/ɔgɐdʰ/ 'Plenty'	/ʧʰɐg/ 'Goat'	Velar	Unaspirated voiced stop
4	ঘ	/gʰ/	/ghɔr/ 'Home'	/ɐgʰɐt/ 'Injury'	/bɐgʰ/ 'Tiger'	Velar	Aspirated voiced stop
5	ঙ	/ŋ/		/kɔŋkɐl/ 'Skeleton'	/bæŋ/ 'Frog'	Velar	Nasal murmur
6	চ	/ʧ/	/ʧɐl/ 'Rice'	/ɔʧɔl/ 'Still'	/ʧɐmoʧ/ 'Spoon'	Post-alveolar	Unaspirated unvoiced affricate
7	ছ	/ʧʰ/	/ʧʰɐgol/ 'Goat'	/biʧʰɐnɐ/ 'Bed'	/mɐʧʰ/ 'Fish'	Post-alveolar	Aspirated unvoiced affricate
8	জ	/ʤ/	/ʤɔl/ 'Water'	/ɐʤkɐl/ 'Now a days'	/kɐʤ/ 'Work'	Post-alveolar	Unaspirated voiced affricate
9	ঝ	/ʤh/	/ʤʰɔɽ/ 'Storm'	/mɐʤʰe/ 'in middle of'	/sãʤʰ/ 'Evening'	Post-alveolar	Aspirated voiced affricate
10	ঞ	/ɲ/		/goʃɐi/ 'vaisnava guru'		Post-alveolar	Nasal murmur
11	ট	/ʈ/	/ʈɔk/ 'Sour'	/ʈɐʈkɐ/ 'Fresh'	/moʈ/ 'Total'	Palatal	Unaspirated unvoiced stop (retroflex)
12	ঠ	/ʈʰ/	/ʈhɐkur/ 'idol'	/kãʈʰɐl/ 'Jack fruit'	/kɐʈʰ/ 'Wood'	Palatal	Aspirated unvoiced stop (retroflex)
13	ড	/ɖ/	/ɖɐl/	/ɐɖɖɐ/		Palatal	Unaspirated voiced stop (retroflex)

(continued)

Table 2.1 (continued)

S. No.	Grapheme	IPA symbol	Example Bangla words with phoneme in			Place of articulation	Manner of articulation
			Initial	Medial	Final		
			'Pulses'	'Gossip'			
14	ঢ	/ɖʰ/	/ɖʰɐl/ 'Shield'	/ɔɖʰel/ 'Plenty'		Palatal	Aspirated voiced stop (retroflex)
15	ণ	/ɲ/		/ʈʰɐɲɖɐ/ 'Cold'		Palatal	Nasal murmur
16	ত	/t/	/tumi/ 'You'	/bɐtɐsʰ/ 'Wind'	/otit/ 'Past'	Alveolar	Unaspirated unvoiced stop
17	থ	/tʰ/	/tʰɐkɐ/ 'Stay'	/kɔtʰɐ/ 'Talk'	/pɔtʰ/ 'Road'	Alveolar	Aspirated unvoiced stop
18	দ	/d/	/doi/ 'Curd'	/kodɐl/ 'Spade'	/pɔd/ 'leg'	Alveolar	Unaspirated voiced stop
19	ধ	/dʰ/	/dhɐn/ 'Paddy'	/ɐdʰɐr/ 'Container'	/bɔdʰ/ 'Kill'	Alveolar	Aspirated voiced Stop
20	ন	/n/	/nɐ/ 'No'	/ɔnek/ 'Many'	/bon/ 'Woods'	Alveolar	Nasal murmur
21	প	/p/	/pɐkɐ/ 'Ripe'	/kɔpɐl/ 'Forehead'	/pɐp/ 'Sin'	Bilabial	Unaspirated unvoiced stop
22	ফ	/pʰ/	/phɔl/ 'Fruit'	/ɐpʰʃoʃ/ 'Repentance'	/sɐpʰ/ 'Clear'	Bilabial	Aspirated unvoiced Stop
23	ব	/b/	/boi/ 'Book'	/kobitɐ/ 'Poetry'	/ʃɔb/ 'corpse'	Bilabial	Unaspirated voiced stop
24	ভ	/bʰ/	/bhɔy/ 'Fear'	/ɐbʰɐsʰ/ 'Hint'	/lɐbʰ/ 'Profit'	Bilabial	Aspirated voiced stop
25	ম	/m/	/mɐ/ 'Mother'	/ɐmɐr/ 'My'	/ɐm/ 'Mango'	Bilabial	Nasal murmur

(continued)

Table 2.1 (continued)

S. No.	Grapheme	IPA symbol	Example Bangla words with phoneme in			Place of articulation	Manner of articulation
			Initial	Medial	Final		
26	র	/r/	/rɔkt̪o/ 'Blood'	/kɔrɐ/ 'Do'	/tomɐr/ 'Your'	Alveolar	Trill
27	ল	/l/	/lɐl/ 'Red'	/bɐlɔk/ 'Boy'	/kɐl/ 'Time'	Alveolar	Lateral
28	শ, ষ	/ʃ/	/ʃɔlɐ/ 'Stick'	/ɐʃi/ 'Eighty'	/pɔlɐʃ/ 'A flower'	Post alveolar	Fricative
29	স	/s/	/sɐt/ 'Seven'	/bæst̪ô/ 'Busy'	/tɐs/ 'Playing cards'	Alveolar	Fricative
30	হ	/h/	/hɐt/ 'Hand'	/bihɐr/ 'To travel'	/bɐh/ 'Wah'	Glottal	Fricative
31	ড়	/ɽ/		/bɔɽo/ 'Big'	/ʃɔɽ/ 'Conspiracy'	Palatal	Unaspirated flap (Retroflex)
32	ঢ়	/ɽʰ/		/driɽʰɔ/ 'Rigid'	/bɐɽʰ/ 'Flood'	Palatal	Aspirated flap (Retroflex)
33	য়	/j/		/pɐjɐ/ 'Leg of a table'	/hɔj/ 'Is'	Palatal	Approximant
34		/w/			/hɐwɐ/ 'Wind'	Bilabial	Approximant

based categorization the palatographs are used for determining the place of articulation, where tongue is the moving articulator. Video capture is used primarily for determining lip closure. The advancement in spectral processing and interpretation has made use of spirometers and EGG important for pathological investigation, though has become replaceable for acoustic phonetic analysis for two obvious reasons. One is that wearing the electrodes is not only inconvenient but also makes the normal informants somewhat tense. Therefore one cannot really examine acoustics of free normal speech. The other is that spectral processing is very reliable, fast and can be done offline even on previously recorded speech.

In the present report Electropalatograph (EPG) in conjunction with video camera has been used to determine the place of articulation of a large number of consonants. For acoustic signatures spectral structure has been used. The EPG system is used to objectively ascertain the place of contact made by the tongue with the hard palate and video camera is used to determine if the closure is made at the lips. For extracting the spectral information commercially available soft-ware packages like Cool-Edit Pro and Wave surfer has been used.

The EPG consists of a custom made palate to fit the informant. This artificial palate extends from the root of the teeth only up to the beginning of the soft palate. The extension of the palate deep into the palate to cover velar region fully is avoided as it would otherwise cause discomfort to the wearer even may do harm to the soft palate. It has 62 contact points (electrodes) on the lower surface where the tongue can make a contact. These arc distributed in eight rows, which correspond to particular articulatory regions as shown in Fig. 2.1. This palate is connected to the EPG electronic instrumentation which produces running frames showing the points by back squares at which a contact is made.

EPG system also records the speech signal and provides both spectral and wave form representation (Fig. 2.2). All information are time-aligned with the EPG frame representation. Figure 2.2 depicts the production of a plosive in V-V context with occlusion, burst and voice onset indicated therein.

For EPG analysis the informant has to wear the false palate in his/her mouth and hold an electrode in one hand. It takes some time for the informant to get

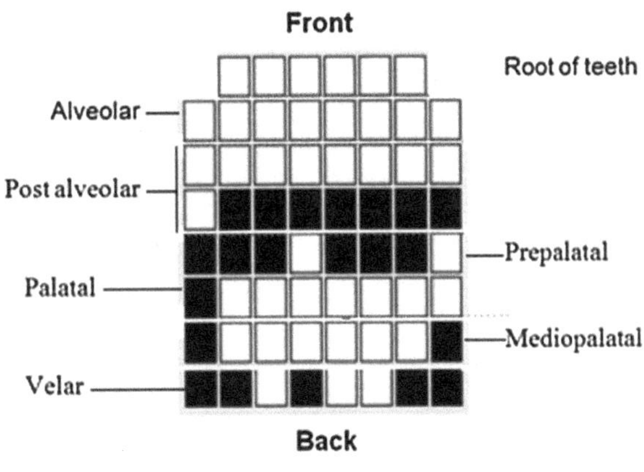

Fig. 2.1 Articulatory regions of the custom made palate

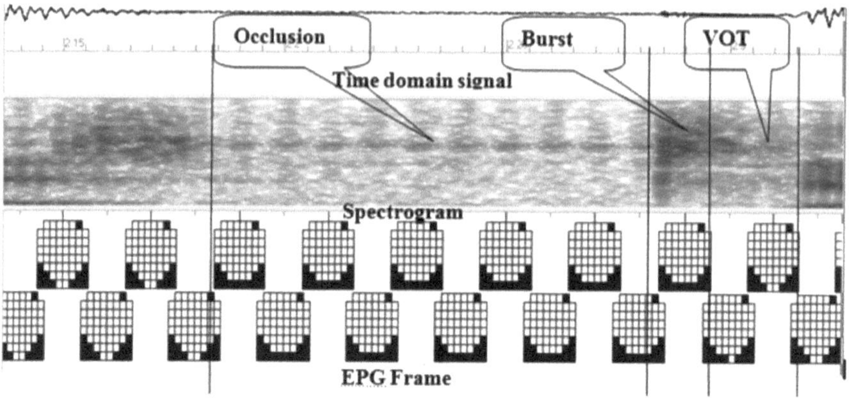

Fig. 2.2 An example of the output from EPG

accustomed. There is no doubt that the system is slightly cumbrous, uncomfortable and highly expensive in the Indian context. However, there being no other more reliable method available for determining the place of contact of tongue and as International Phonetics Association recommends this instrument for assuring standardization of data this has been used for the present study.

2.3 Speech Material

For acoustic analysis of the manner of articulation of consonants, data in VCV context was taken from the speech corpus of DIT, Kolkata [www.cdackolkata.in]. Altogether speech data of 5 speakers (2 male and 3 female) had been considered for this study. All of them are native speakers of Standard Colloquial Bangla. They were selected through a properly constituted selection test. The nativity was assured using three criteria, (1) both the parents must belong to the relevant districts, (2) the informant was reared by their parents, and (3) the education up to the higher secondary level was through Bangla medium. The metadata of the informants is given in Table 2.2. The words used were taken from a phonetically balanced word-set (PBW) of about 600 words. The PBW was prepared such that as far as

Table 2.2 Metadata of the informants

Speaker	Age (years)	Sex
1	25	Male
2	36	Male
3	25	Female
4	36	Female
5	42	Female

Table 2.3 Number of different VCV segments

Type of VCV segments	No. of segments			
	Velar	Retroflex	Dental	Bilabial
Unvoiced unaspirated plosive	256	276	263	242
Unvoiced aspirated plosive	260	151	228	69
Voiced unaspirated plosive	206	49	186	200
Voiced aspirated plosive	117	28	110	65

practicable all necessary CVC combinations used in the language are available and there occurrences are nearly the same. These words were embedded in a neutral carrier sentence for reading and recording.

Furthermore natural words containing the possible CC cluster and the C at three different positions were also added to the list. For the analysis of voiced consonants only those VCV segments were considered in which the occlusion period shows distinct voice bar, and similarly for unvoiced ones only those VCV segments having no voicing in the occlusion region were considered. The number of different VCV segments for different consonants was given in Table 2.3. For the analysis using EPG the unit list consists of a set of nonsense VCV sequences in which V represents the seven Bangla vowels /u/, /o/, /ɔ/, /a/, /æ/, /e/ and /i/ and C represents all the consonants of Bangla. For stops all VC sequences for the 20 stop consonants were added in the list.

It may be seen that for unaspirated voiced and aspirated voiced retroflex plosives the number of data were remarkably low. In Bangla the occurrence of these consonants in VCV context is somewhat rare. This is also observed in case of aspirated unvoiced and aspirated voiced bilabial plosives. For analysis which required EPG the appropriate recording material was read out by one male and female informant, wearing the respective artificial sensor palates. Only one informant of each sex was used because of logistic problem as the palates cannot be made in India. The EPG dynamic contact information was recorded in the PC at a regular interval of 10 ms. The acoustic signal from a microphone supplied with the instrument was recorded at 22,050 sampling frequency 16 bits mono PCM format, synchronized with EPG frames. The reading is limited for a maximum of 20 min duration in a session for the convenience of the informants. Ten repetitions each of the VCV and the VC from the aforesaid list were recorded for each speaker for the determination of place of articulation of consonants.

2.4 Methodologies

The objective analysis of consonantal sounds can have two approaches, one is primarily instrumental and the other uses signal processing techniques to extract acoustical information from digitized audio signals. As has already been mentioned, instrumental approach is not often preferred because it affects the naturalness of free speech. It is reported that the dynamics of certain parameters like formant frequencies show strong correlation with the place of articulation (Datta and Mukherjee 2011) and is largely used in speech technology applications (Das Mandal 2007; Choudhury 2006; Datta 1988; Datta et al. 1978a, b, 1981; Datta and Ganguly 1981, 1985) for the categorization of plosives/affricates with respect to the place of articulation. It may also be noted that these parameters can be reliably extracted, using appropriate signal processing techniques, from online as well as recorded data and therefore can represent some characteristics of free speech. However the instrument EPG may be used to validate the description of the place of articulation obtained through signal processing techniques. For various reasons, mainly methodological, the categories show overlap significant enough for this approach for resolution of ambiguities. The EPG provides possibly the best way to determine the place of articulation with a reasonable degree of accuracy. Moreover the lacuna regarding the naturalness may be somewhat addressed by allowing the informant to get accustomed with the contraption. It is hoped that with some practice this disadvantage is mitigated to a large extent.

Therefore in the present book the primary emphasis is given on signal processing techniques as far as acoustic parameters are concerned. For the place of articulation the EPG is used.

2.4.1 Acoustic Analysis

The signatures for different manners of production related to voicing and aspiration are available in both the signal domain and spectral domain presentation. Plosives and affricates of Bangla (*spɔrʃɔbɔrɲɔ*),[1] like all other languages of the Indo-Aryan family, are traditionally presented in a matrix form where the columns represent manners and rows places of articulation. These show a five-way contrast in terms of both manner and places of articulation. In Table 2.4 these are organized in a 5×5 matrix where the places of articulation are arranged from velar to labial. The categories mentioned in this table are traditional subjective views. Of these the second row presents the affricates the others are plosives/stops. The third row presents the retroflex plosives.

The acoustics of all the consonants are presented below separately for individual plosives, affricates, fricatives, lateral, trill, and taps. An additional section on consonantal murmurs has been included though it is not a separate phoneme, yet it exhibits characteristic acoustic signatures and requires special attention in speech technology applications. These represents voiced occlusion portions of plosives and affricates. The reason behind inclusion of the section is that one does not find in-depth study of acoustics of these portions in literature. It may not be out of order here to mention that the definition of the so called unit 'phoneme' is not really a

[1]Literally meaning phones produced by touching.

Table 2.4 Traditional matrix representation of Bengali plosives and affricates

	Unaspirated unvoiced	Aspirated unvoiced	Unaspirated voiced	Aspirated voiced	Nasal
Velar	[k]	[kʰ]	[g]	[gʰ]	[ŋ]
Palatal *(affricate)*	[ʧ]	[ʧʰ]	[ʤ]	[ʤʰ]	[ɲ]
Post-alveolar *(retroflex)*	[ʈ]	[ʈʰ]	[ɖ]	[ɖʰ]	[ɳ]
Alveolar	[t]	[tʰ]	[d]	[dʰ]	[n]
Labial	[p]	[pʰ]	[b]	[bʰ]	[m]

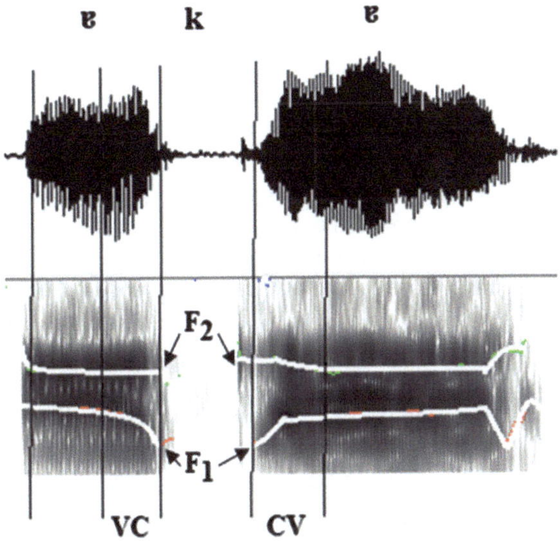

Fig. 2.3 Illustration of phoneme boundaries

crisp one. In fact, in real speech it is hard to locate them accurately. Let us consider a simple vowel-plosive-vowel syllable (Fig. 2.3). We can locate the burst and the voice onset time (VOT) as well as the occlusion period quite confidently. But that is not the whole plosive.

The plosive extends into the vowel, so called VC and CV segments, only then it becomes cognitively meaningful. Again the whole voiced, the quasi-periodic region namely from the beginning of CV to the end of the signal in the above figure, is not the vowel alone. If one listens to only this whole quasi periodic signal segment one will hear the CV syllable. For normal listener the difference on including the preceding and the succeeding consonants is not even noticed. Again one has to listen only to the complexity-wise steady part of the signal to hear the pure vowel. Even then the boundaries of the steady part are also fuzzy, at least cognitively. When one studies the acoustics of speech sound, one must give adequate

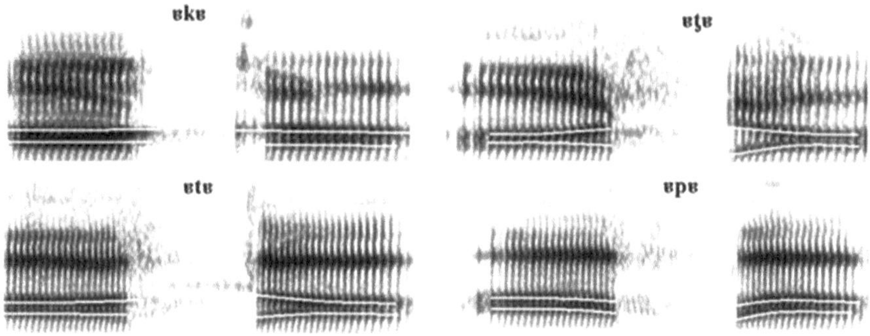

Fig. 2.4 Illustrating transitional cues for place of articulation of plosives

importance to the cognitive domain and be careful about segmentation so that relatively confident zones are selected for parameter extraction.

In the following sections acoustics primarily related to the manners of production will be discussed. It is because the other basis of categorization, namely the place of articulation, does not normally put any distinctly robust imprint on the spectral structure of the different elements of the consonantal segments. The consonantal segments constitute occlusion of the vocal tract, burst, release and segment of aspiration (see Fig. 2.4). Its effect is strongly marked on the quasi-periodic signal representing the adjoining vowel with which the consonants interact strongly. In fact the parameters like amount and direction of formant transition and the time of transition (Datta et al. 1981; Datta and Ganguly 1981), are reported to be distinctively related to the place of articulation. Figure 2.3 illustrates this with a particular VCV syllable. It may be mentioned here that the transition of the second formant provides the best cue (Datta and Mukherjee 2011) for this purpose as this formant represents the backwardness of the tongue hump. One may notice from Fig. 2.4 that the transition of F_1 and F_2 are different for different consonants. While F_1 transition is always downwards towards the burst, since at the point of release tongue always touches the palate and F_1 is related to the height, F_2 movement can be both upwards and downwards. The F_2 transition varies normally according to the place of articulation of the consonant. There is no F_2 transition for [k] in [ɐkɐ] since it has almost the same place of articulation as that of the vowel.

2.4.2 *Acoustics of Plosives*

Plosive sounds result from the blocking of the vocal tract by the tongue or lips, allowing the air pressure to build up behind the closure, and then by a sudden release of it. This mechanism produces sounds like /p/, /k/ etc. The time for which the tract remains blocked is called occlusion period. In Bangla for each place of articulation there are five manners of production. Of these four manners related to voicing and aspiration are produced by a unique process of co-ordination between the glottis and the articulators namely the tongue or lips. Nasals are produced by connecting the nasopharynx with the oral cavities by the opening of the velum. The aforesaid co-ordination may be explained with reference to Figs. 2.5, 2.6, 2.7 and 2.8. Each

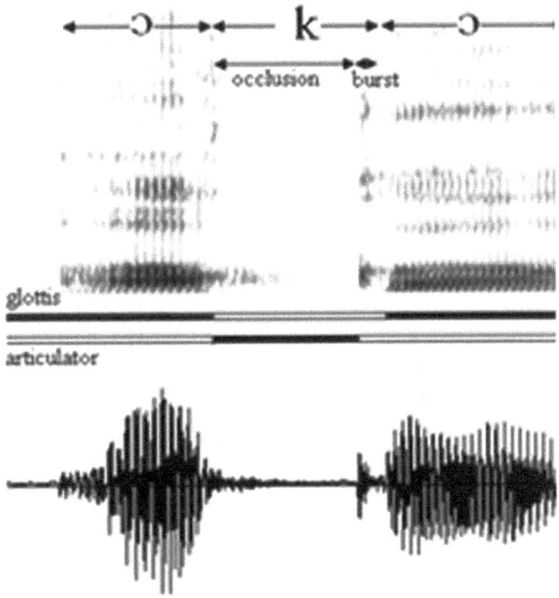

Fig. 2.5 Mechanism for production of unaspirated unvoiced plosives

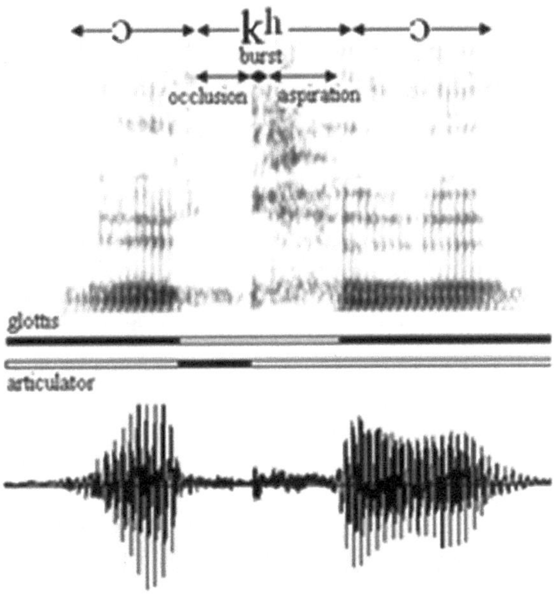

Fig. 2.6 Mechanism for production of aspirated and unvoiced plosive

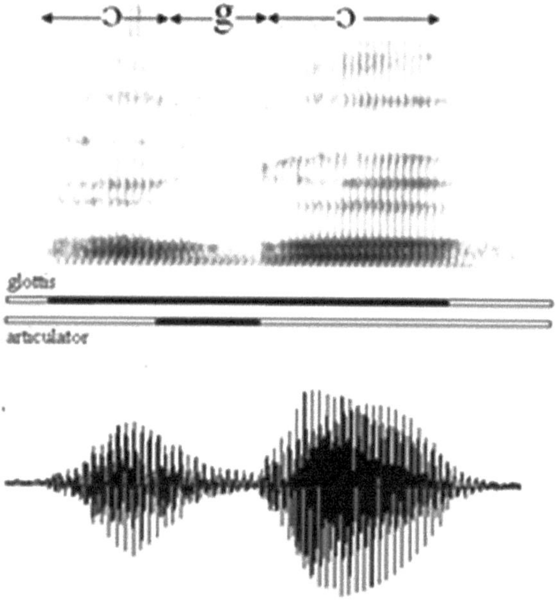

Fig. 2.7 Mechanism for production of unaspirated and voiced plosive

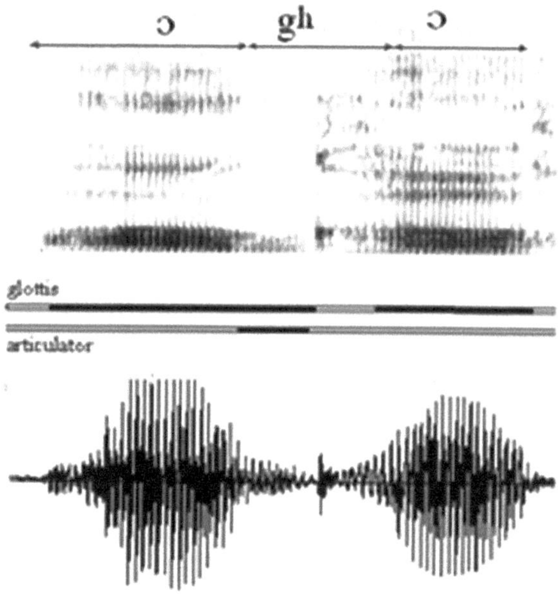

Fig. 2.8 Mechanism for production of aspirated and voiced plosive

figure has three sections. The uppermost and the lowermost sections show respectively the signal and the corresponding spectrogram. The two bars in the middle section are used to indicate the open (no colour) and the close (black colour) status of the articulator and the glottis. Closed glottis produces the source voice and the closed articulator suppresses energy in the higher frequencies. The mechanism for the production of different manner of articulation is explained as follows:

(1) When both glottis and articulators are closed voice bars indicative of the voiced manner of plosive are produced.
(2) If the articulator remains closed but the glottis is open it indicates a voiceless occlusion.
(3) When both the articulator and the glottis remain open aspiration is produced because of the turbulence created through the rushing of air through glottis. (One may note that in this particular context the glottis may not be fully open.)
(4) If along with the condition (1) above the velum is open connecting the nasal cavity, nasal murmurs are produced.
(5) Closed glottis with open articulator produces normal quasi-periodic sounds.
(6) If along with (5) velum remains open these give rise to the nasal set of vowels.

Voice Onset Time (VOT) is used in this article to define the time taken to close glottis after the release of closure of the oral tract by the articulator. When the length of VOT is large enough to make it acoustically perceptible, one hears them as the aspirated counterpart of the phonemes even if there is not much aspiration energy. Again if the VOT is small (say <20 ms), even if there is significant aspiration energy the aspiration is not heard.

2.4.3 Acoustics of Affricates

An affricate is a combination of plosive and fricative. In Bangla there are again four types of affricates according to their manner of production, (1) unaspirated unvoiced affricates /ʧ/, (2) aspirated unvoiced affricates /ʧ^h/, (3) unaspirated voiced affricates /ʤ/ and (4) aspirated voiced affricates /ʤ^h/. As we have seen in the case of plosives in last section these manners also reflect unique co-ordination between the opening and closing of the articulators with those of the glottis. However for affricates the important thing to note is that unlike plosive the release is not explosive in nature. The release is rather slow allowing the articulator to make a constriction for some time before the full opening. During the period of constriction air passing through it produces turbulence to cognitively reflect as noise. This together with the occlusion has the cognitive sense of affricates. This part is indicated by the grey in Figs. 2.9, 2.10, 2.11 and 2.12. On the other hand a fricative (as will be discussed in a later section) does not have an occlusion period and the frication is significantly longer. In voiced affricate the occlusion is voiced whereas in unvoiced affricates this part is silence. The spectral structures of the frictional area may be seen from the corresponding fricatives discussed in the next section.

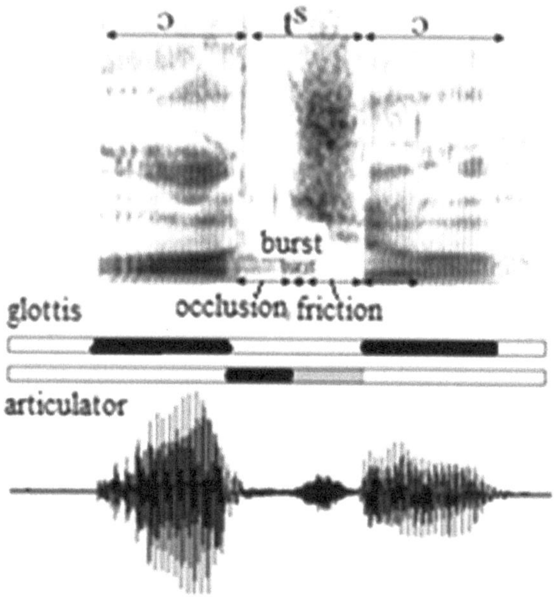

Fig. 2.9 Mechanism for production of unaspirated and unvoiced affricate

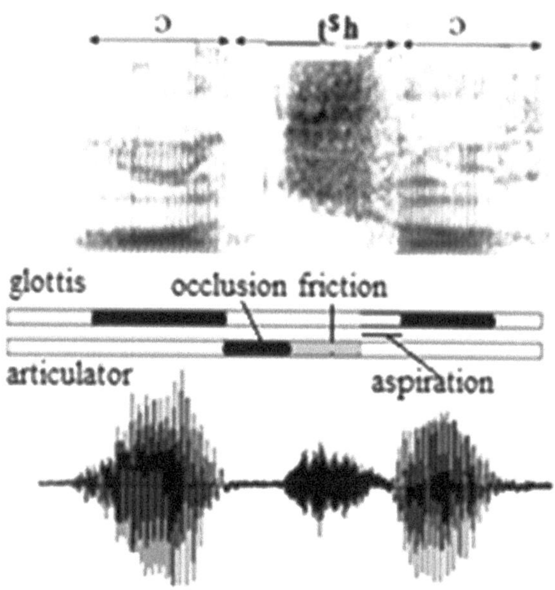

Fig. 2.10 Mechanism for production of aspirated and unvoiced affricate

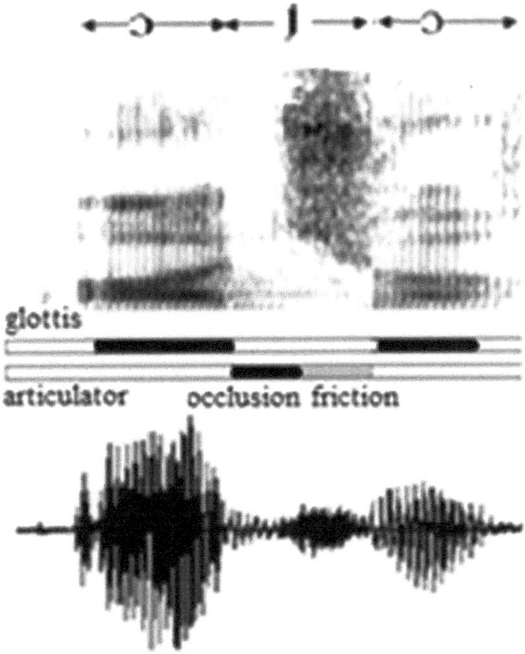

Fig. 2.11 Mechanism for production of unaspirated and voiced affricate

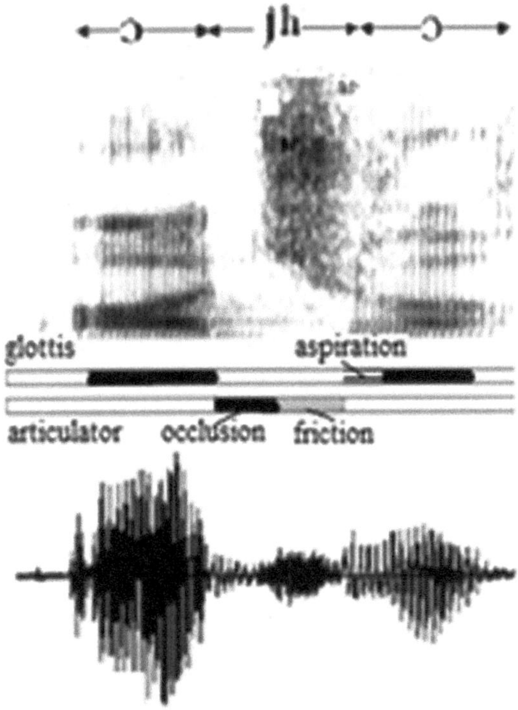

Fig. 2.12 Mechanism for production of aspirated and voiced affricate

2.4.4 Acoustics of Fricatives

Fricative consonants are distinguished from other speech sounds by their manner of production and the fact that they are noisy in character because of their predominant randomness in the spectral structure. While the signal of the voiced speech is quasi-periodic in nature all fricatives are quasi-random in Bangla. Bangla does not have any voiced fricative. The Bangla fricatives are produced by the flow of air through a narrow constriction produced in some region in the vocal tract. The major articulators in Bangla which form the constriction are usually, the tongue blade and the lips. This noise is then filtered by the vocal tract, with the acoustic cavity in front of the constriction contributing the greatest influence in filtering. Figure 2.13 presents the mid-sagittal section of the vocal tract for different place of articulation of the fricatives.

Fricatives are often divided into two groups: sibilants and non-sibilants. Sibilants refer to a cognitive property, a hissing noise, which is the primary way to distinguish between alveolar fricatives and dentals. There may be two different mechanisms for fricative sound generation: (1) an obstacle source, where the turbulence may be generated at a rigid body approximately perpendicular to the airflow, or (2) a wall source where sound is generated along a rigid wall parallel to the flow. For /s, z/ and /š, ž/ the teeth are considered to be the obstacles while the upper lip is an obstacle parallel to the flow for /f, v/, /z, ž, and v/ (Shaddle 1990). These are not Bangla fricatives.

Fricative Spectra, according to the theory of speech production and simplified electrical models (Heinz and Stevens 1961), can be characterized by poles and zeroes, which depend on the location of the constriction which is also the source of excitation. Also the resonance frequencies are inversely related to the size of the front cavity. It appears that the back cavity would act as the shunt for the source of noise and therefore their effect on the noise spectra would be revealed as anti-resonances.

There are three fricatives [ʃ], [ş] and [s] in Bangla. The basic spectral character of these unvoiced fricatives relate to pure random noise without harmonic structure. The major characteristic spectral features are the resonances and anti resonances. Another important feature is the spectral tilt. Figures 2.14, 2.15 and 2.16 represent

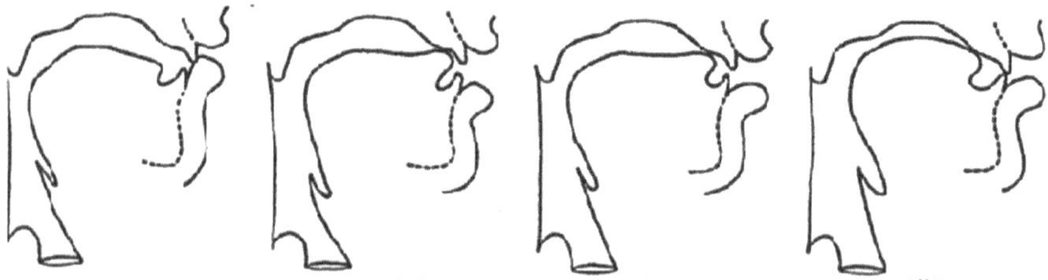

Fig. 2.13 Midsagittal cross-section of a vocal tract constriction (Perkel 1969)

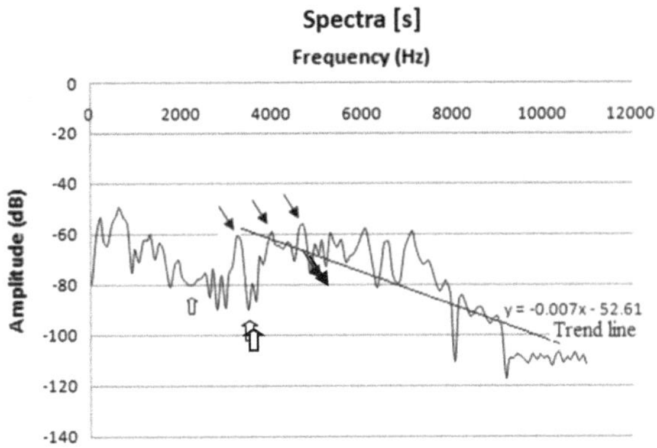

Fig. 2.14 Example of spectra of alveolar fricative of Bangla

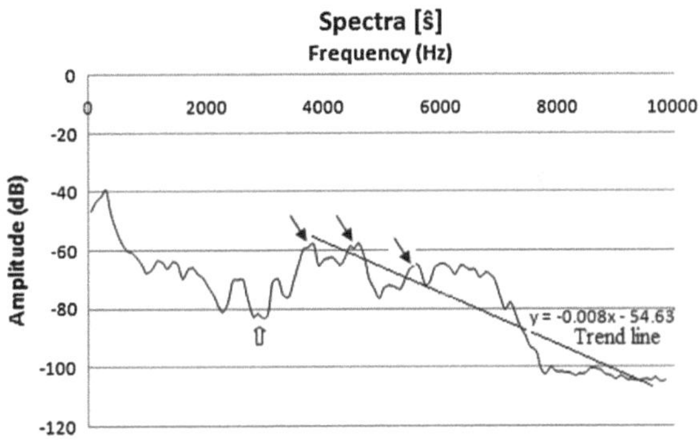

Fig. 2.15 Example of spectra of retroflex fricative of Bangla

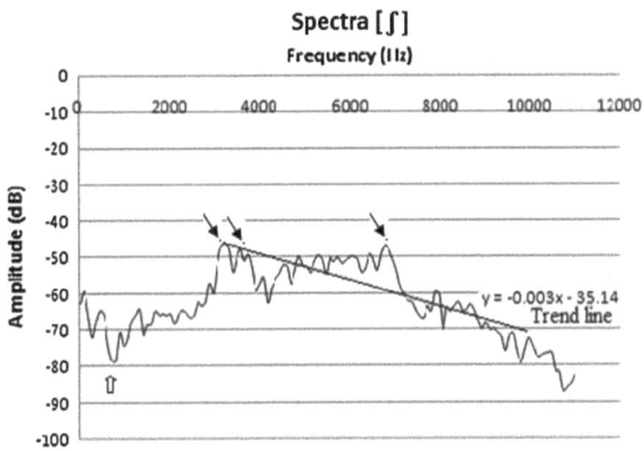

Fig. 2.16 Example of Spectra of palate-alveolar fricative of Bangla

examples of the spectra of the three fricatives in Bangla. There is always a major anti-resonance in the low frequency region of the spectra. This will be referred to as the trough hereafter. In these figures it is indicated by a vertical wide arrow. Only the major peaks after the trough are considered for further description and these are indicated by single inclined arrows. Spectral tilt is considered for characterizing

Table 2.5 Statistics for duration, F_1 and F_2 for Bangla taps and trills

Phone	Statistics	Duration	F_1	F_2
r	Mean	51.5	499.5	1423.0
	SD	14.6	150.7	405.7
ɽ	Mean	83.9	353.8	1006.2
	SD	12.1	12.1	305.4
ɽh	Mean	116.2	359.2	1058.8
	SD	13.0	31.3	314.1

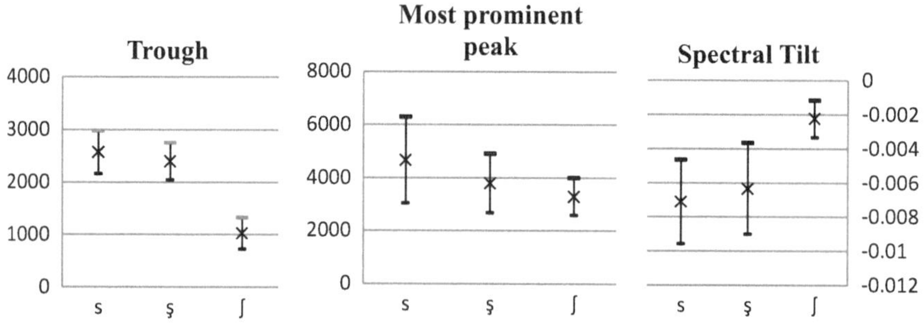

Fig. 2.17 Relative positions of [ʃ], [s] and [ʂ] in three acoustic feature spaces

only the high energy part of the spectrum starting with the first peak point as shown by the trend lines in the figures. The coefficient of 'x' in the equation for trend line is taken as the measure of a tilt.

The acoustic features need to be examined are: locations of (1) the trough, (2) the first three peaks, (3) the most prominent peak and (4) the slope of the spectral tilt. The statistics of these acoustic parameters are presented in Table 2.5. Except the tilt all values are in Hertz. For tilt the figures represent the slope. It may be seen from the table that the frequencies of trough, the most prominent peak and the slope of the tilt clearly distinguishes [ʃ] from the other two fricatives. However these parameters are not much distinctive for [s] and [ʂ]. In fact this is vindicated by Fig. 2.17, which represents pictorially the separability of these parameters for the three fricatives. The mean value is indicated by the cross and the lines represent the spreads of the features. The spread equals mean ± SD. In fact the frequencies of the presented peaks as well as the spectral tilt each makes [ʃ] disjoint from both [s] and [ʂ].

Standard deviations of the frequency of the trough as well as those for peaks for the three fricatives are quite small indicating good stability. There is a small problem though; the peaks do not really represent the mean values of different formant frequencies as they do in vowel spectra. Usually in the spectral data the first, the second or the third peak may appear at any frequency whatsoever.

One could approach to estimate the formants from the probability distribution of the spectral peaks appearing in all the signals of a particular fricative. Figure 2.18 presents these distributions for the three fricatives. It may be seen from these

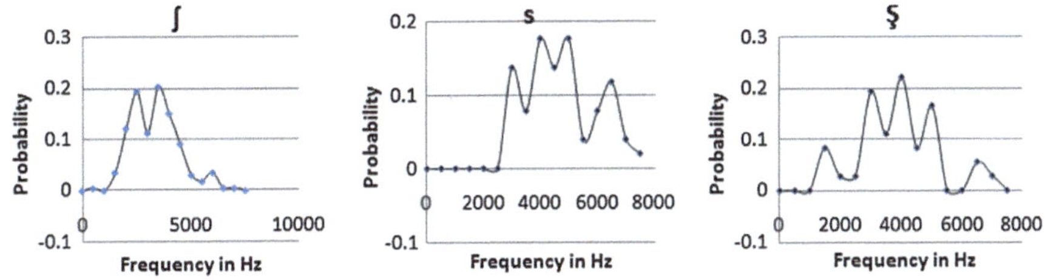

Fig. 2.18 Probability distribution of peaks in spectral energy

diagrams that for [ʃ] two formants at 2500 and 3500 Hz are quite prominent. For [s] the first formant is at 3000 Hz. Three other formants are at 4000, 5000 and 6500 Hz. Similarly for [ş] these are respectively at 1500, 3000, 4000 and 5000 Hz. The bins used for calculating the distribution was of 500 Hz width. The aforesaid values represent the mode values not the mean values. One may note that [ş] is characterized by a peak appearing at a relatively low frequency of 1500 Hz.

2.4.5 *Acoustics of Laterals*

The lateral sounds are produced when the mid-coronal closure is complete and the air escapes around one or both sides of the closure. Bangla has only one lateral sound /l/. We shall see later that this could be both bi-lateral and unilateral. For these sounds the air passage is fully blocked medially but at someplace below the dental region an escape route exists some times on both sides (bi-lateral production) and at other times only one side (unilateral production). Because of the fact that the obstruction is at the dental region, the back cavity is large. This should have reflection in the first two formant frequencies. Furthermore the obstruction of the air passage is likely to cause relatively larger reduction of energy in the higher bands though a general loss of energy of the laterals compared to the adjoining vowels is expected. Figure 2.19 represents the spectrgram of a /VLV/ syllable with white line representing the F_0. The lateral can be easily identified here.

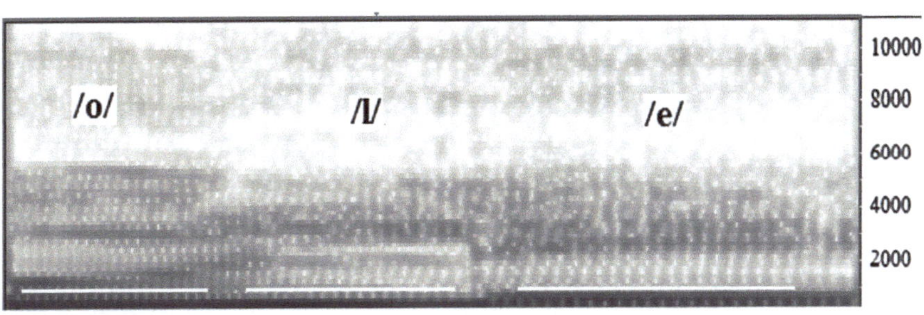

Fig. 2.19 Spectrogram of a /VLV/ syllable

Figure 2.20 gives the spectral sections of the three sections. One could easily see the reduction of the first harmonics.

As mentioned earlier the medial obstruction is likely to create reduction of signal energy. To see whether this happens, it is decided to examine the loss of energy in the low frequency band as well as the high frequency band. For this F0 is determined first and then this value is used as the cut off frequency for band pass filters provided in Cool-Edit Pro (Fig. 2.21)

Figure 2.22 presents the loss of energy on lateralization separately for the low frequency band (F_0) and the high frequency region. There is loss in 96% of the

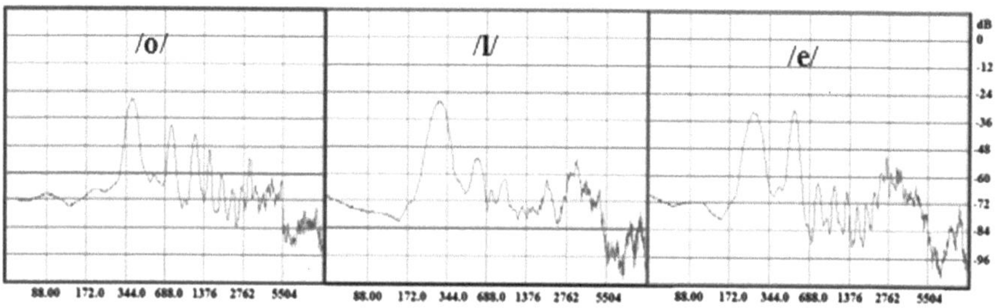

Fig. 2.20 Spectral sections of constituent phonemes

Fig. 2.21 Low pass filter band

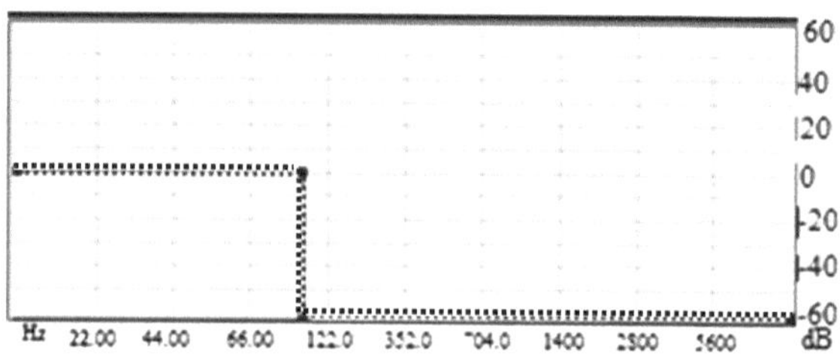

Fig. 2.22 Loss of energy in dB on lateralization

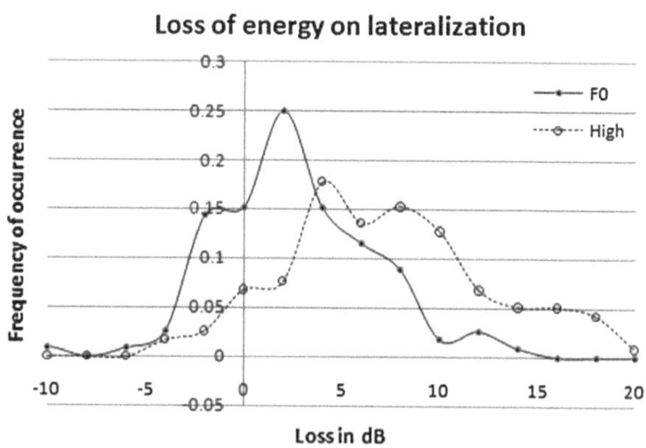

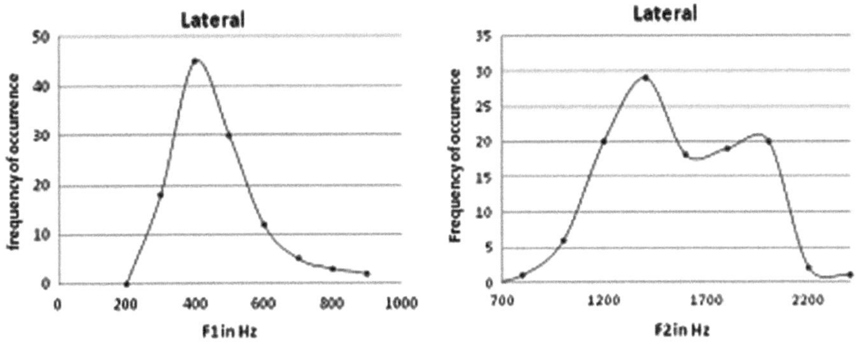

Fig. 2.23 Distribution of laterals in F_1 and F_2 axes separately

Fig. 2.24 /l/ in the F_1-F_2 plane

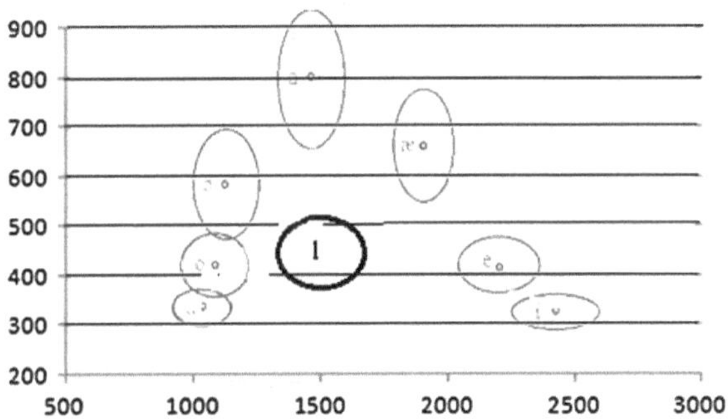

cases for the laterals in the high frequency band and 81% cases in the low frequency band, the average loss being about 7 dB and 4 dB respectively. Thus there is in general loss of energy in both the bands with that in the F_0 being more prominent.

Figure 2.23 shows the distribution of laterals separately for F_1 and F_2. The mean value of F_1 of the laterals was found to be 421 Hz with the standard deviations of 142 Hz. The distribution of F_2 being far from normal mean value may not be used for representation. One may note that the distribution is bi-modal for the second formant. These two modes are due to the influence of the front-back position of the preceding vowel. The first mode is located at about 1400 Hz and the second mode at about 2300 Hz.

Figure 2.24 shows the position of [l] with reference to Bangla vowels in the F_1-F_2 plane. It may be noticed that the position of the lateral is in the region on inner side of the relatively free space unoccupied by the boomerang shaped vowel distribution. It indicates the possibility of distinguishing the lateral using the first two formant frequencies in case of Bangla.

2.4.6 *Acoustics of Trills and Taps*

Trills may be described as phones generated by the vibration of some supralaryngeal articulators (tongue-tip, uvula, lips). This vibration is caused by the aerodynamic forces, whereas taps and flaps involve active muscular movements of the tongue. The setting up of the tongue-tip vibration involves muscle contraction, shape and elasticity requirements of the tongue, and a sufficient pressure difference across the constriction. Once trilling is initiated, tongue-tip vibration is maintained as a self-sustaining vibratory system. Pressure builds up behind the lingual constriction until it forces the tongue-tip to open. The air rushes out causing the pressure to drop. The reduction of pressure along the constriction due to the Bernoulli Effect makes the tongue-tip spring back to the contact position. This cycle repeats (Catford 1977; Ladefoged 1967; Spajić et al. 1996; Mcgowan 1922). For trills the tongue body is more constrained than for the tap. Trills coarticulate less with neighbouring vowels. The acoustic waveform for voiced trills exhibits a low amplitude murmur during the tongue-tip closure.

Tap is produced with a single contraction of the muscles so that one articulator (usually the tongue) is thrown against another articulator. The tongue body is less constrained than in trill. The tap is articulated with a restricted short apico-alveolar closure and more pre-dorsum lowering than other alveolars (Recasens and Pallare 1999).

It is said that Bangla has one trill [r] one unaspirated retroflex tap [ɽ] and one aspirated retroflex tap [ɽh]. Figure 2.25 shows examples of spectrograms and wave forms of two instances of [r]. The tap positions are indicated by the arrows. The left one shows a single tap and the right figure shows two taps indicating a trill. The thick horizontal lines above the wave forms indicate the duration of [r]. Similarly Fig. 2.26 shows corresponding examples for [ɽ] and [ɽh].

Bangla [r] is traditionally reported to be a trill, which means there would be more than one contact of the tongue with the palate. Similarly [ɽ] and [ɽh] is traditionally reported to be taps indicating a single contact. However acoustic examination of a large number of instances of multiple speakers from the aforesaid data base revealed that only 10% of the instances for [r] exhibited a double contact the rests

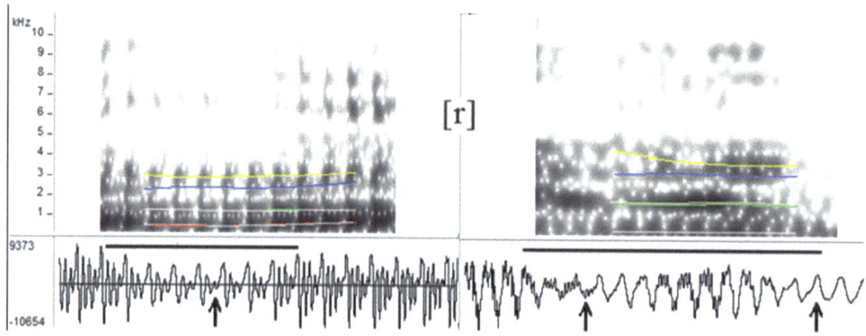

Fig. 2.25 Examples of spectra and wave forms of [r]

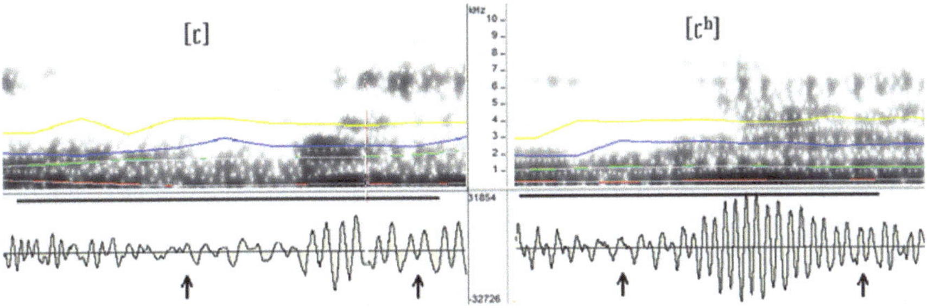

Fig. 2.26 Examples of spectra and wave forms of [ɽ] and [ɽh]

only one contact. This indicates that, with Bangla native speakers, [r] is generally a tap. The same examination with [ɽ] revealed that all instances have a single contact indicating it to be a tap. However 88% of [ɽh] exhibited double contacts indicating that this may be a trill or a rolled utterance.

Since for all these sounds the oral cavity is divided by the tongue the position of contact is likely to influence the formant frequencies, particularly the F_2. As the data base consists of these sounds in various different vowel contexts the co articulation effect may be quite significant. For locating F_1 and F_2 frequency distribution of the formants has been drawn (Fig. 2.27).

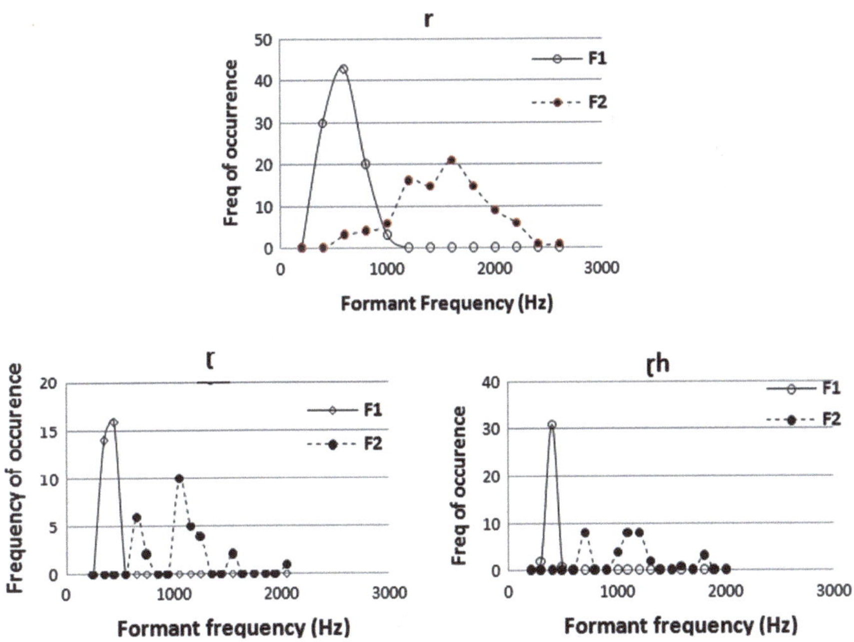

Fig. 2.27 Frequency distribution of F_1 and F_2 for the taps and trills

The distributions for F1 are unimodal indicating absence of influence of adjoining vowels. Which is expected, as in all cases the tongue touches the palate. Modes observed from Fig. 2.27 for F_1 are 600, 450 and 400 Hz respectively for [r], [ɽ] and [ɽh]. The low value for [ɽ] and [ɽh] indicate possible lowering of the jaw which may be necessary to accommodate curling of the tongue. The frequency distributions for F_2 exhibit multi modality with major modes at 1200 and 1600 Hz for [r], 650 and 1050 Hz for [ɽ], and 700 and 1100 Hz for [ɽh]. The multi-modality indicates contextual influences of adjoining vowels. So does large values of standard deviations.

Table 2.5 presents the mean and standard deviations for duration, F_1 and F_2 for these three consonants. The SDs are generally low for F_1 indicating almost no influence of the adjoining vowels. However the comparatively larger values of standard deviations for F_2 are expected as the distributions are far from normal. A reference to Fig. 2.27 reveals that the distribution of F_2 is multimodal. In fact this puts into question the effectiveness of mean and standard deviation to characterize the variability of this particular parameter. The examination of the detail data revealed strong influence of the vowels on F_2. The standard deviations of durations for all these consonants are fairly low. Along with that the mean values being quite different, duration may considered as an inter class distinctive feature for these consonants. Figure 2.28 shows the position of the taps and trills in F1-F2 plane in relation to Bangla vowels represented by gray ovals. The vowel positions are taken from an earlier report (Datta 1988).

Figure 2.29 presents frequency distribution of duration for these consonants. Modes are seen to be well separated. It may be seen that the mode of duration for [r] is around 45 ms, and is distinctively less than those for [ɽ and ɽ^{h}]. The retroflexing of the tongue is expected to need more time. The modes for the last two are at 105 and 120 ms respectively. The large value for ɽh is expected as this trill generally contains two separate contacts.

It needs to be mentioned that there are only few words in Bangla which contains [ɽh]. These are /driɽhɔ/, /muɽhɔ/ and /ɐʃɐɽh/, /gɐɽhɔ/, /guɽhɔ/, /mɐɽhi/, and /ruɽhɔ / respectively meaning 'firm', 'dumb', 'the name of a month', 'thick, 'secret' 'gum' and 'rude'. It is very likely that in common speech [ɽh] may not, and even need not, be pronounced in its proper manner for general communication purposes.

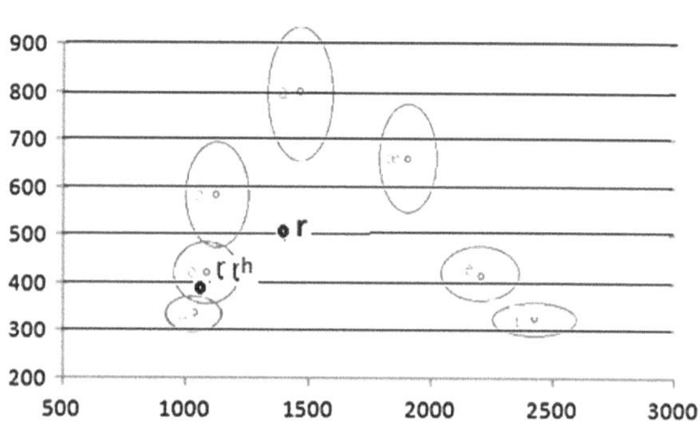

Fig. 2.28 Position of Bangla taps and trills wrt vowels

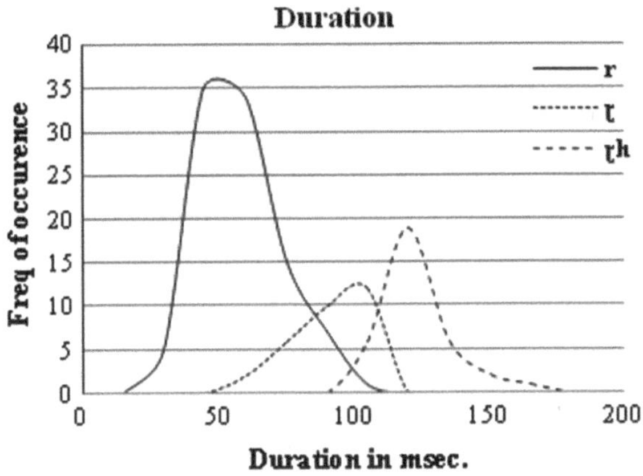

Fig. 2.29 Frequency distribution of duration of [r, ʈ and $ʈ^h$]

2.4.7 Consonantal Murmurs

Consonantal murmurs are defined here as speech sounds produced by the occlusion of the oral tract when the vocal chords are oscillating. These are not phonemes by themselves but plays an important role in manner distinction and therefore demands investigation. These may be purely oral, e.g. [g, g^h, d, d^h, ɖ, $ɖ^h$, b, b^h, ʤ, and $ʤ^h$] in Bangla or purely nasal, e.g., [n, ɲ, ɳ, ŋ, m] in Bangla. These murmurs need not be confused with that of murmured vowels, a typical feature in many Indo-Aryan languages (Chap. 1, Sect. 1.5), wherein murmur is associated with a lax glottis introducing aspiration energy into the voice source (Klatt and Klatt 1990; Hillenbrand et al. 1994; Stevens 2000; Ladefoged and Antananzas-Barroso 1985). Though these consonantal murmurs constitute a large portion of consonantal sounds not much attention towards the acoustic structure of these sounds seems to be evident. Major attentions appear to be given to the anticipatory influence of these sounds on the duration of the previous nucleus vowel and to the cognitive aspects of these influences. The objective here is to study only the segmental acoustics of these murmurs.

The articulatory mechanism for the production of these sounds may be understood from the schematic diagram presented in Fig. 2.30. The air stream from lungs causes oscillation of the vocal folds in the closed glottis during the production of the preceding vowel. The murmurs are produced by the occlusion of the vocal tract while the glottis remains closed and therefore voicing source remain active. If during this occlusion the velum also remains closed the oral murmurs are produced. The sound that emerges is not direct but by radiation across the walls of the mouth cavity. However, when the velum is open sound also leaks through the nasopharynx with associated addition of nasal formants.

The closure is likely to cause primarily a severe reduction of the sound energy. One would expect the loss to be primarily in the high frequency region as radiation allows low frequency components to come out albeit with some absorption. As the

Fig. 2.30 Schematic of the articulatory configuration for consonantal murmurs

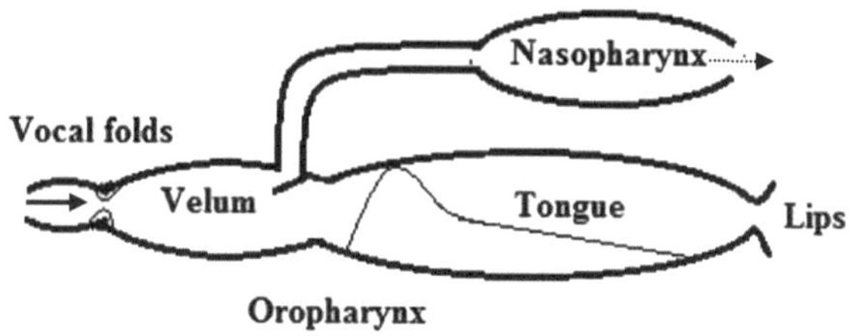

closure takes place after the production of preceding vowel the air-stream velocity across the glottis is likely to start fading out because the sink is closed. This may result in the lowering of F_0. Furthermore this closure may also cause a back pressure disturbing the normal free functioning of the vocal folds, the signature of which is likely to be found in the amplitudes of the first two harmonics, H_1 and H_2, relative to that of F_0. With these premises in view the study is conducted with about 200 murmurs of which 89 are oral the rest are nasals. A comparative study of the following parameters is done for the consonantal murmurs with respect to the preceding vowels:

(A) Total energy
(B) Energy in the low frequency band
(C) Change in the F_0
(D) Amplitude of $H_1$and H_2 with respect to that of F_0
(E) The relative amplitudes of H_1 and H_2.

The VC segments have been manually extracted from the speech corpus of CDAC, Kolkata. The simple affirmative sentences spoken by five male and five female native informants (see Table 1.1 of Chap. 1) has been used for the purpose. The number of oral murmurs and that of nasal murmurs are 89, and 104 respectively. Table 2.6 represents the number of samples for murmurs with different places of articulation.

For the purpose of extracting parameters the signals corresponding to each sentence need to be segmented. This was done manually through an examination of both the signal profile and the corresponding 3D spectrogram.

Figure 2.31 presents the spectrogram and corresponding waveform of a simple affirmative sentence, 'ʃe nɐbiker nɐm kɔlombɐʃ', meaning '*that sailor's name*

Table 2.6 Distribution of samples across murmurs

Murmur	No. of samples
[b]	32
[g]	18
[d]	17
[dz]	21
[m]	26
[n]	87

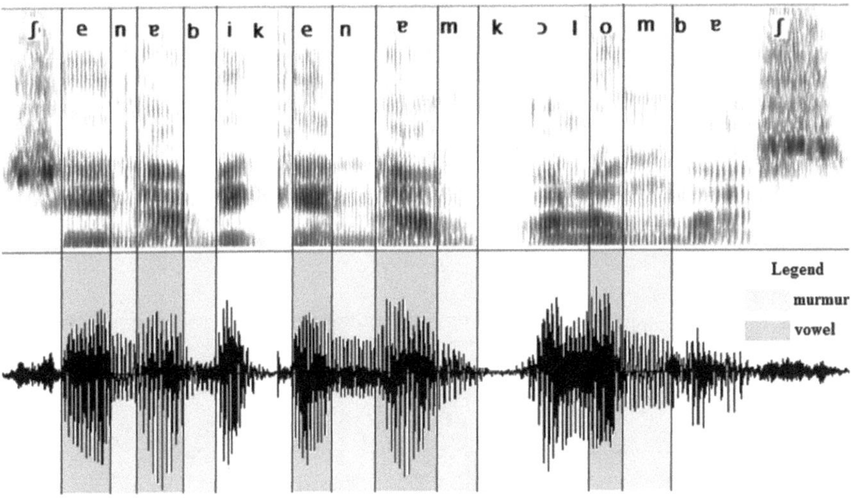

Fig. 2.31 Example illustrating segmentation of the sentence /ʃe nɐbiker nɐm kɔlombɐʃ/

Fig. 2.32 Spectrogram of vowel murmur pair

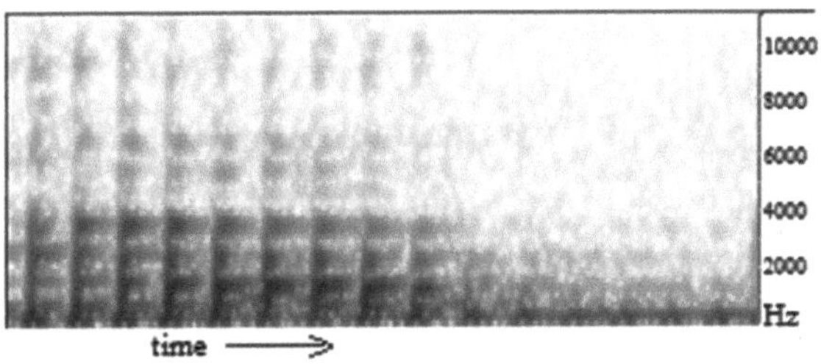

(is) *Columbus*'. The segmentation of vowels and succeeding consonantal murmurs are marked therein.

The following procedure is adopted generally for the extraction of the parameters. For this segments of signal of length 40 ms are selected on both sides bordering the start of closure of the vocal tract. This closure generally leaves clear identifiable visible signature in both the forms of representation (see Fig. 2.31). However sometimes in both the forms the signature is not equally prominent. For example in a vowel-murmur combination with the vowel [u] or [i] the amplitude profile may not provide a clear separability, in these cases 3D form is helpful. Cool-edit Pro is used for the purpose. Figure 2.32 represents an example of the spectrogram of a typical vowel murmur combination.

Figure 2.33 represents the corresponding spectrogram when the higher energy above F_0 is filtered out using the filter package provided in Cool-Edit Pro. The average loss of total energy due to the closure ^{T}E for the same window of 40 ms is measured using the following simple formulae:

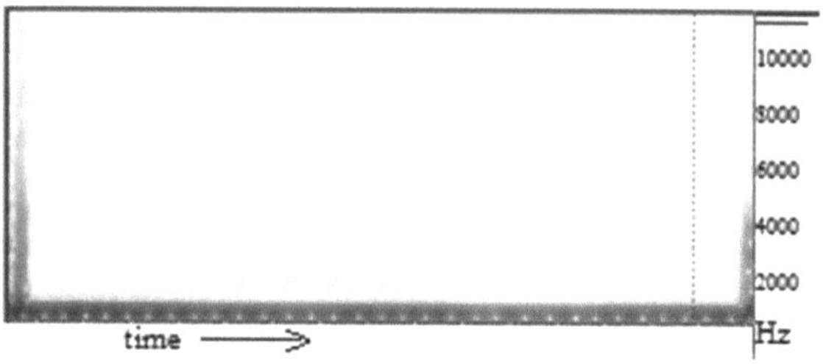

Fig. 2.33 Spectrogram of the same pair after low pass filter

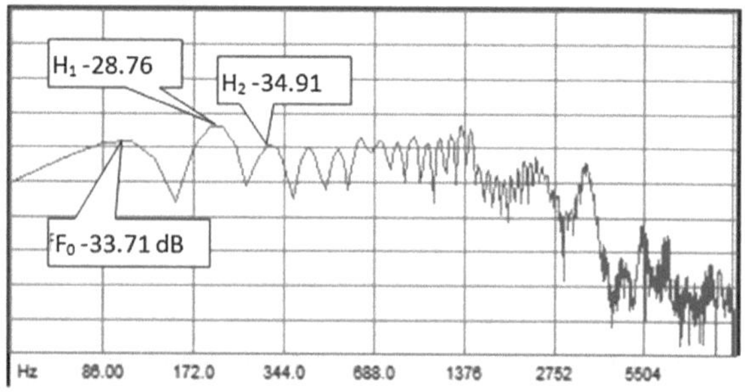

Fig. 2.34 Spectral section showing measurement of amplitudes of F_0 and H_1 and H_2

${}^{T}E = ({}^{T}E_V - {}^{T}E_M)/{}^{T}E_V$ and ${}^{L}E = ({}^{L}E_V - {}^{L}E_M)/{}^{L}E_V$, where ${}^{T}E_V$ and ${}^{T}E_M$ represent the average of 40 ms of signal taken respectively just before and just after the closure. Also ${}^{L}E$ presents the proportional loss of energy in the low frequency band just covering the fundamental due to the closure and ${}^{L}E_V$, ${}^{L}E_M$ represents the average energy of 40 ms of the filtered signal taken respectively just before and just after the closure.

The spectrum section presented in Fig. 2.34 may be used to describe the measurements of the fundamental frequency F_0 and the amplitudes of F_0, and of the first two harmonics H_1 and H_2. The change in F_0 due to the closure is measured in semitone using the formula:

$\Delta F_0 = 12 * \log_2 ({}^{v}F_0/{}^{m}F_0)$, where ${}^{m}F_0$ and ${}^{v}F_0$ presents respectively the fundamental frequency of the murmur and that of the preceding vowel.

The sudden closure of the air passage at the mouth cavity is likely to cause accumulation of air coming from the lungs. This may develop a back pressure affecting the vocal fold oscillations which in turn may cause variation in the amplitudes of H_1 and H_2 with respect to that of F_0. It is necessary to see if there is any such consistent effect. For this, the amplitudes of F_0, H_1 and H_2 are measured from the average spectrum sections for 40 ms of the vowel signal just before the transition and that of the murmur signal just after the transition. Three differences namely ${}^{A}F_0 - {}^{A}H_1$, ${}^{A}F_0 - {}^{A}H_2$ and ${}^{A}H_1 - {}^{A}H_2$ were computed both for the vowel segment and the murmur segment, where ${}^{A}F_0$, ${}^{A}H_1$, and ${}^{A}H_2$ are amplitudes of F_0, H_1, and H_2 respectively.

The occlusion of the vocal tract has been premised here to cause significant loss of energy in the output signal. Figure 2.35 presents the distribution of the proportional losses (PL) of total energy and of low energy due to closure separately for

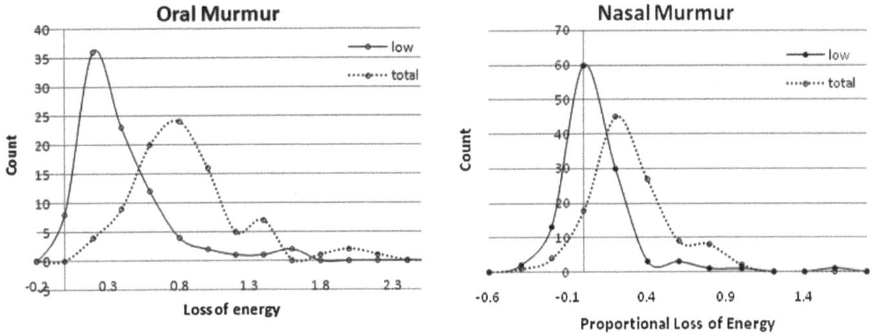

Fig. 2.35 Distribution of proportional loss of energy due to closure

the oral and nasal murmurs. The PL of total energy is seen for all oral murmurs and 96% for the nasal murmurs. Thus as expected, there is generally a loss of total energy for the murmurs. As regards to the PL of energy in the low frequency band just covering the F_0 we see that while there is a loss for all oral murmurs a small but significant portion of about 15% of nasal murmurs do not show any such loss. This may be due to the fact that sound leaks through the nasopharynx with associated introduction of nasal resonances. The mode values are taken to indicate the amount of PL instead of the average values simply because the distributions appear to be far from normal. While with total energy the modal value of PL for oral murmurs is 0.8 dB that for nasal is 0.2 dB. The low value of PL for nasals may be again due to the presence of a parallel path for the sound through the nasopharynx

It is interesting to note that the modal value of the loss in the energy of the fundamental frequency due to closure is 0.8 dB for oral murmurs whereas there is no loss for nasal murmurs. This is because for nasals there is a clear output path through nasopharynx. It seems to indicate that of the total output of energy in the case of oral murmurs the bulk of low energy is through radiation from the walls and that the bulk of the energy directly out of the mouth opening consists of the high frequency band. It must be remembered that the figures given above are proportional loss; actual loss would be much larger as these would by multiplied by the energy of the preceding vowel.

Figure 2.36 presents the change ΔF_0 of fundamental frequency due to the closure of the oropharynx. Again we take the largest mode value instead of average for the reasons stated earlier. The largest mode value of 0 semitones indicates there is generally no change in the F_0 due to closure. However a closer perusal of Fig. 2.36 indicates that about 60% of oral murmurs and 67% of nasal murmurs do indicate a decrease of F_0. If we consider only samples which indicates a decrease then the average value are 3.8 semitones for oral and 5.4 semitones for nasal murmurs.

This decrease is expected as the air stream velocity across the glottis is expected to fall with the closure of the oropharynx. However under the same expectation on the air stream velocity one would expect the leakage through nasopharynx would work in opposition to the said effect of the closure. One should therefore expect

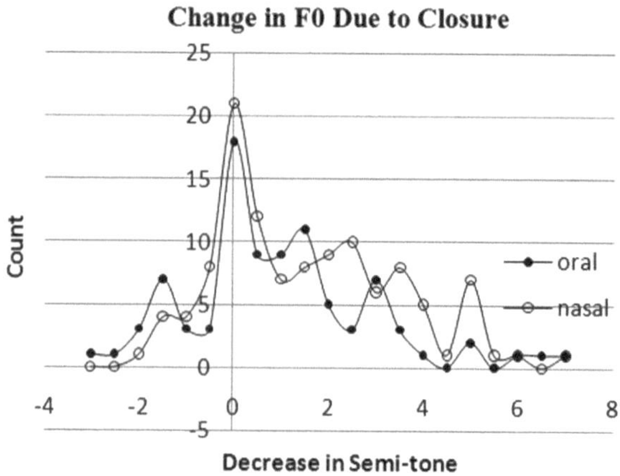

Fig. 2.36 Distribution of ΔF_0 due to closure of the oropharynx

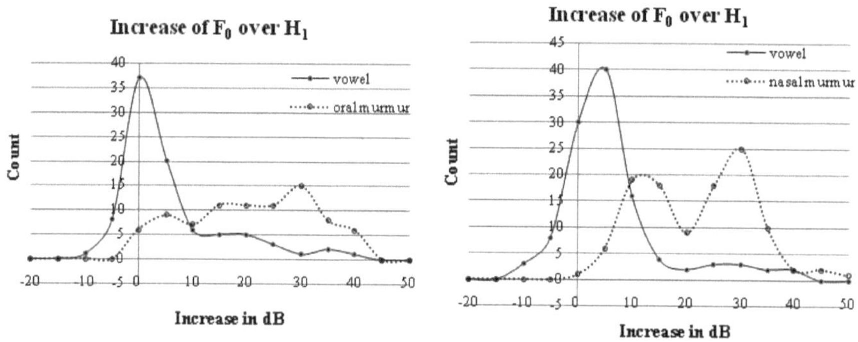

Fig. 2.37 Distribution of samples with the amount of increase of F_0 over H_1

lower value of the decrease for nasals instead of the observed significantly higher values. The main cause of the decrease may however be related to the functioning of the laryngeal oscillators which needs direct observation of the vocal cord movements when the oropharynx is closed and there is a consequent back-pressure generated by accumulated air in the front cavity on the glottal structure.

Figure 2.37 shows the distribution of the decrease of the amplitude of H_1 over that of F_0 for both the vowel segments and the murmur segments. In the case of oral murmurs the modal value of decrease is 0 dB for the vowel and 30 dB for the murmur. However about 48% of the vowels in these pairs showed some decrease while 90% of the murmurs show a decrease of the amplitude. Again for the vowel-nasal murmur pairs modal value of decrease is 5 dB for vowels and that for the murmur is again 30 dB. Results seem to indicate clearly that murmurs cause significant decrease in the amplitude of H_1 over that of the fundamental.

Figure 2.38 shows the distribution of the decrease of the amplitude of H_2 over that of F_0 for both the vowel segments and the murmur segments. In the case of oral

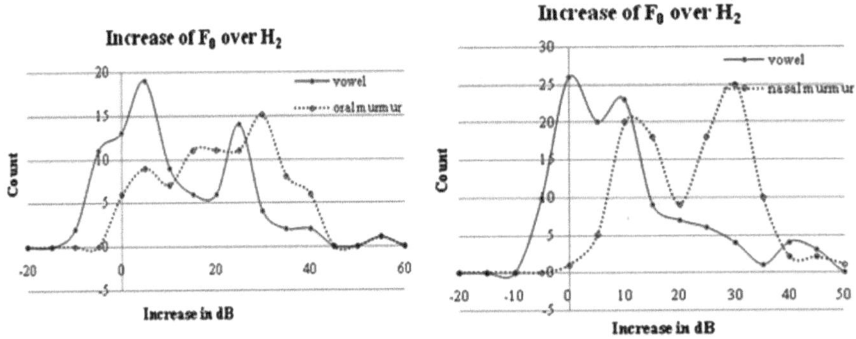

Fig. 2.38 Distribution of samples with the amount of increase of F_0 over H_2

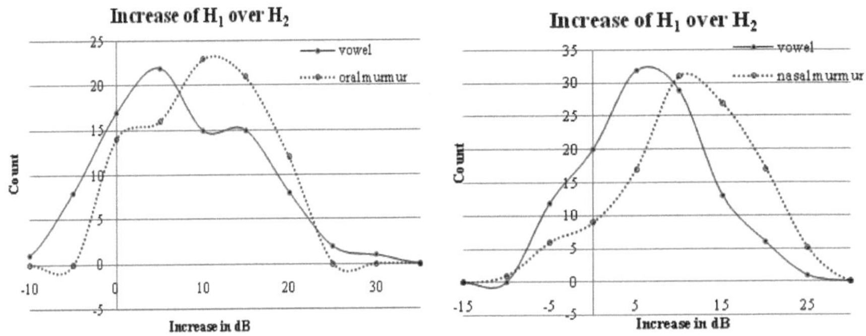

Fig. 2.39 Distribution of samples with the amount of increase of H_1 over H_2

murmurs the modal value of decrease is 0 dB for the vowel though an additional mode of almost equal strength is observed at the modal value of 10 dB. That for the murmur is at 30 dB same as that for H_1. About 68% of the vowels in these pairs showed some decrease while 98% of the murmurs show a decrease of the amplitude. Again for the vowel-nasal murmur pairs modal value of decrease is 5 dB and that for the murmur is again 30 dB. Results seem to indicate clearly that murmurs cause significant decrease in the amplitude of the second harmonic over that of F_0.

Figure 2.39 presents the distribution of the increase of the amplitude of H_1 over that of H_2 for both the vowel segment and the murmur segment. In the case of oral murmurs the modal value of increase is 10 dB for the vowel. That for the murmur is at 15 dB. About 71% of the vowels in these pairs showed some increase while 81% of the murmurs show an increase of the amplitude. Again for the vowel-nasal murmur pairs modal value of increase is 5 dB and that for the murmur is again 15 dB. About 71% of the vowels in these pairs showed some increase while 81% of the murmurs show an increase of the amplitude.

An interesting observation from Figs. 2.37 and 2.38 is that nasal murmurs show very distinctive prominent bi-modal distributions. This led to an investigation to see if they have any relationship with the place of articulation or even with the preceding vowel class. Careful investigation revealed no such correspondences. In fact this bimodality remained even for the two major murmur classes, m and n, as well as for isolated vowel groups. The energy loss due to closure is significant. The loss is most for the higher frequencies, total for oral murmurs but insignificant for nasal murmurs. The loss of energy for F band is negligible This indicates that in speech the energy output in the very low frequency band is primarily through radiation. The decrease of F_0 frequency has been observed in a large number of cases. A decrease in the average value is 3.8 semitones for oral and 5.4 semitones for nasal murmurs which is unexpected because for nasal a separate path for airflow exists. This put into doubt the conjecture that the decrease is due to the decrease in the velocity of the airflow across vocal folds. The cause of decrease may be related to the functioning of the glottal oscillations due to back pressure generated by the accumulated air in the oral cavity. This gets strong support when we examine the amplitudes of H_1 and H_2 with reference to that of the fundamental. The closure of the vocal tract causes significant decrease in the amplitude of the first and second harmonic with respect to that of F_0. This indicates the possibility of the effect of closure on the functioning of the vocal folds. One may note here that the closure causes accumulation of air in the back cavity which may cause this.

2.4.8 Summary

While summarizing the aforesaid findings on acoustics of Bangla consonants, it may be pertinent to observe that a comprehensive documentation of acoustics of consonants in any of the Indo-Aryan spoken language is hard to find. Also the author could not find data on acoustic properties of the murmurs associated with voiced plosives and affricates in other language group also. At this moment, therefore, it is not always possible to have an inter-lingual comparison. A crisp summary would therefore be in order here. One important aspect of the plosives and affricates in Bangla, in fact for most of the Indo-Aryan speech sounds, is that they have full repertoire of manner classes. The relation of the acoustics of these manners namely the voicing, the aspiration and the nasalization with the unique process of co-ordination between the glottis, the articulators namely the tongue or lips and the velum are described in detail. It is known that a fully closed glottis produces source voice, while an open glottis, in the present context, produces aspiration and an open velum produces nasality. Different interaction between these three produces the five manners namely, unvoiced-unaspirated, unvoiced-aspirated, voiced-unaspirated, voiced-aspirated and nasal.

For fricatives spectral features are examined with reference to (a) locations of the trough, corresponding to the antiresonance of the cavities prior to the source, (b) first three peaks and the most prominent peak corresponding to the resonances of the cavities, post to the source and (c) to the slope of spectral tilt. The frequencies of the presented peaks as well as the spectral tilt shows [ʃ] to be distinct from both

[s] and [ʂ]. On the other hand [ʂ] is characterized by a peak appearing at a relatively low frequency of 1500 Hz.

Acoustics of the only lateral /l/ is influenced by the medial blockage and lateral escape for the air-stream. There is, in general, compared to the preceding vowel a loss of energy in both the F_0 band and the high energy band with that in the F_0 band being more prominent. The mean value of F1 and F_2 of the laterals were found to be 421 and 1502 Hz respectively. The respective standard deviations are 142 and 365 Hz. The position of this lateral in the vowel diagram is quite distinct and is in the inner space of the 'boomerang' shaped distribution of Bangla vowels. One may note that the distribution is bi-modal for the second formant. These two modes are due to the influence of the front-back position of the preceding vowel.

The acoustic investigations reveal Bangla native speakers [r] is generally a tap. This is in contradiction to the traditional description of [r] as a trill. The same examination with [ɽ] revealed that all instances have a single contact indicating it to be a tap instead of its traditional categorization as a trill. We shall see corroboration of these finding later with EPG analysis. However 88% of [ɽh] exhibited double contacts indicating that this may be a trill or a rolled utterance. The determination of the first two formants helps in fixing their position in the vowel diagram. It is found that while [r] has a distinct position, [ɽ, ɽh] are not separable and their positions are congruent with that of vowel [o].

Consonantal murmurs, a new class introduced for studying their acoustics, are defined here as speech sounds produced by the occlusion of the oral tract when the vocal chords are oscillating. As expected, there is generally a loss of about 80% in total energy and only 30% in the F_0 band. This indicates the most of the output energy in the F_0 band is indirect by radiation through the walls of mouth cavity. For nasal murmurs there is, in general, no loss in F_0 band and only about 30% loss in the total energy. This is also expected as there is an additional pathway for the sound through the nasopharynx. As regards to the effect on F_0 it is found that about 60% of oral murmurs and 67% of nasal murmurs do indicate a decrease of F_0. The examination of amplitudes of fundamental and the first two harmonics reveal that the amplitudes of the first two harmonics decrease significantly in comparison with those for the preceding vowels.

2.5 EPG Analysis

While acoustics can provide very distinctive description of manners of consonantal categories, it does not provide useful distinctive robust characteristics for the categories related to place of articulation. Apart from subjective categorization, which has been most prevalent traditionally, there are various objective methods of determining the place of articulation. These include cine-fluorography, ultra-sonography, videography, endoscopy and palatography. The cine-fluorography being medically hazardous it is not used now a day. Ultra-sonography has not yet reached the necessary technological level to provide the needed robustness while endoscopy is very uncomfortable and restricts severely the freedom of the speaker. Thus video capture and palatography remains the main tools in use. The latest development in palatography is the EPG system.

While general principle of using EPG has been briefly described in Sect. 1.2 it is necessary to give some details of experiment conducted for the present study. One issue is the selection of appropriate frame for a particular category of consonant. The other is the methodology used for obtaining a representative frame from a set of frames coming from the repetitions of the same syllable/word. This exercise is necessary for standardization so that the data from different languages are comparable.

Locating the frame or frames for investigating an event must also be well defined. This is necessary as the final data would be a representative of a number of repetitions, and the sequences of each representation should conform to the constituent events in best possible way. An occlusion is easily identified when the frames have the sequence of continuous black cells connect the two sides of the palate showing a complete separation of the front and back halves of the palate (Fig. 1.1 in Sect. 1.2). A closure is indicated when such a frame is preceded by a frame with a discontinuity. A break in this continuity is characteristic of the beginning of a release. The frame characterizing such a break must be seen in the continuity of frames to know whether it is a release or a closure. Contact area is measured by the number of contacts made to the total number of electrodes in the whole EPG frame and is expressed as percentage.

The place of articulation of the friction is determined by the EPG frame where the opening is narrowest. Figure 2.40 shows an example illustrating the procedure for fricatives. Double headed arrow indicates the segment corresponding to the

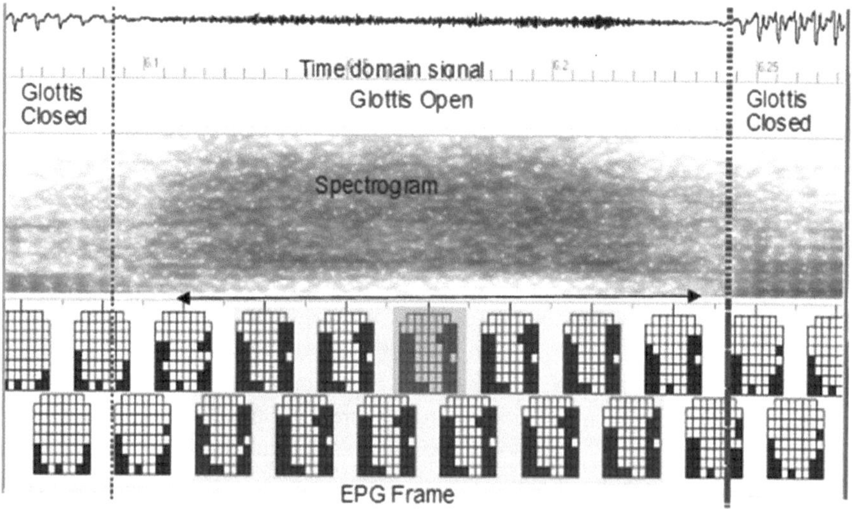

Fig. 2.40 Choice of frame representing the place of articulation for fricatives

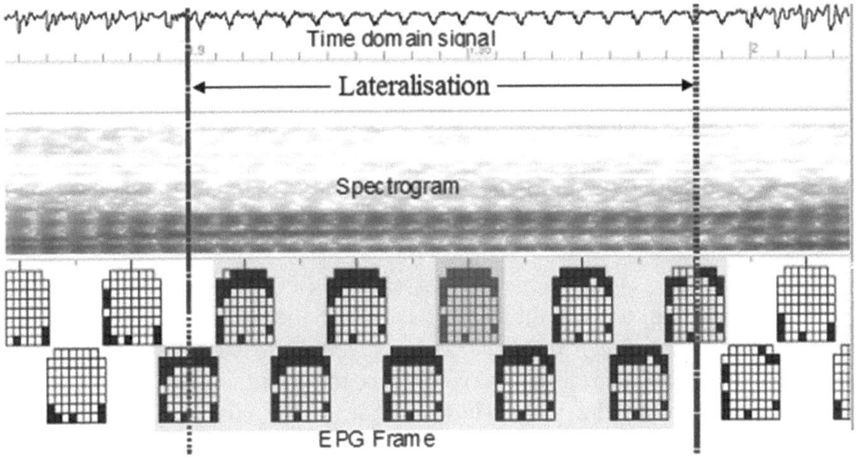

Fig. 2.41 Choice of frame representing the place of articulation for lateral

acoustic event friction. Slightly grayed area represents the frames corresponding to narrowest constriction. The grayer frame at the middle of the friction zone is the selected frame.

Figure 2.41 shows a sequence of EPG frames containing a lateral. The complete medial closure is clear with a large unilateral opening at the right side for this particular instance can also be clearly seen.

For each event like release, closure, constriction etc., related to a phoneme, there are EPG frames the number of which depends on the group being considered, e.g. for individual speaker and a specific context this number is minimum of 10 repetitions. One has to decide on a procedure for getting a frame which shall represent the set of these frames. The issue of registration of the sequence has already been discussed. As indicated in Sect. 1.3 an EPG frame presents an array of small rectangles either black or white. A binary matrix is first obtained by putting '1' for black cell indicating a contact and '0' for white indicating no contact. There are 10 such frames for each event in a sequence, a particular CV or VC context, for each informant. The representative frame is constructed in the following manner. The value of each cell is computed by adding the number of 1's for all these 10 repetitions. Then cumulated values are divided by the number of frames obtained for the speaker for that context. The computed value is used to determine the gray value of the cell where a purely black one indicates that all frames show a contact has been made at that cell. A purely white cell indicates that no contact has been made there. Thus the original binary picture of the frame is converted into a representative gray level picture to accommodate variations in pronunciation.

2.5.1 Plosive/Stop

Figure 2.42 presents one series of EPG frames for VCV utterance where C is an unvoiced unaspirated plosive to exemplify selection of frames to study the place of articulation as plosive and stop for these consonants. The lighter gray shaded frames represents the occlusion period. The deeper gray frames are the terminal ones to be used for determining the place of articulation of plosives and stops.

ক [k]

Figures 2.43 and 2.44 presents the representative EPG frames averaged over the selected frames from ten repetitions, of the release for [k] for each of the seven Bangla vowels respectively for female and male speakers. A general feature that is observed in all such frames is that the contacts begins at the side much earlier going maximally forward to the mid-sagittal position which represents the place of articulation. This indicates a funnel shaped tongue configuration. It can be seen from these that a complete closure is not observed for many vowels due to an inherent restriction of the artificial palate. As mentioned earlier the artificial palate is

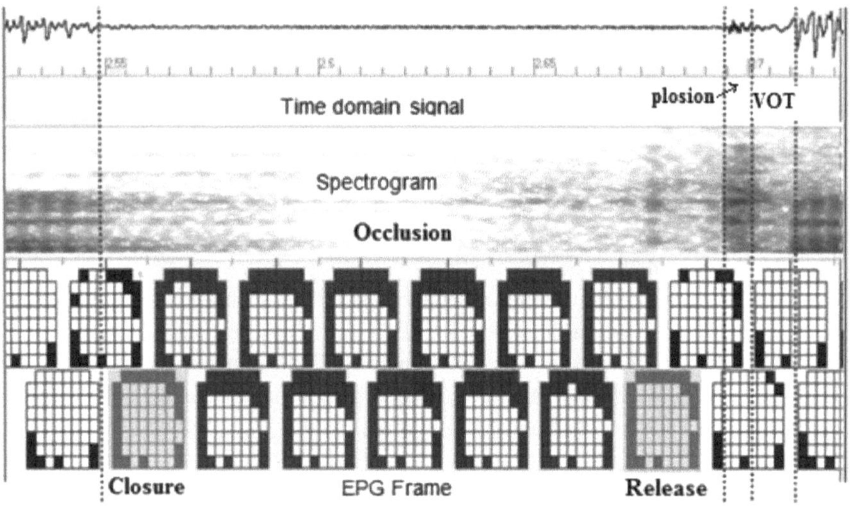

Fig. 2.42 Selection of closure and release frames for plosives

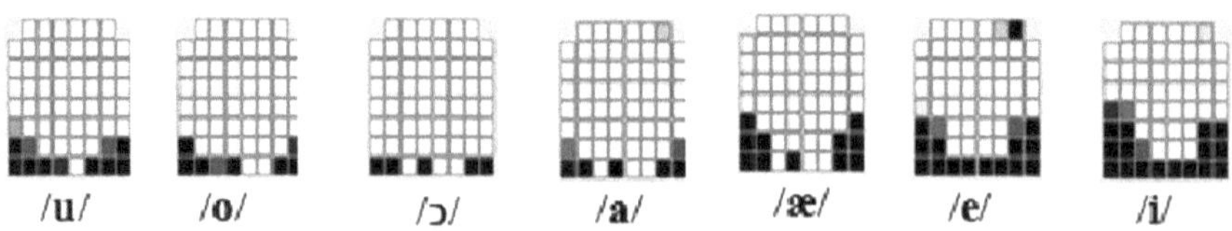

Fig. 2.43 Place of release of [k] for seven Bangla vowels for the female speaker

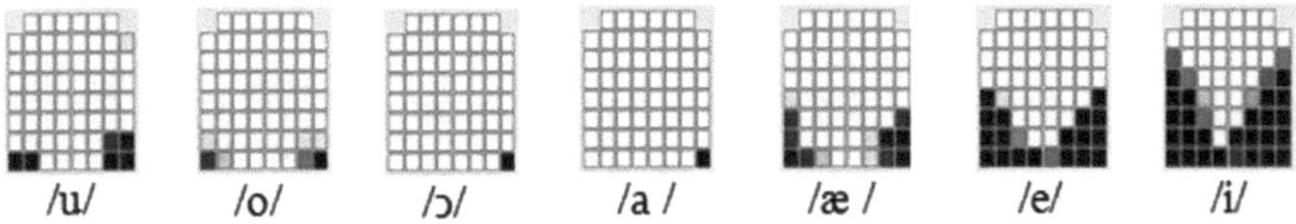

Fig. 2.44 Place of release of [k] for the all Bangla vowels for the male speaker

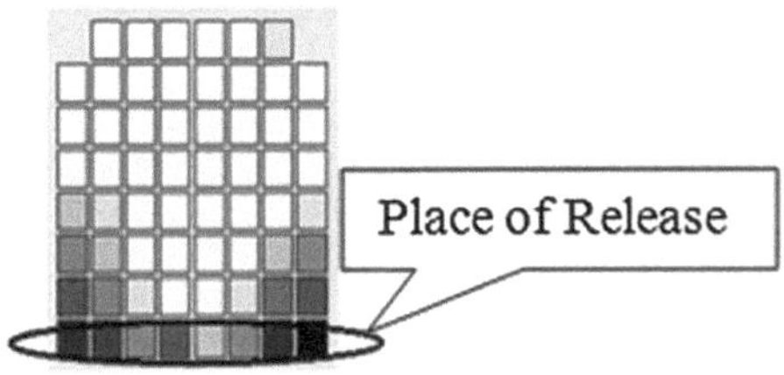

Fig. 2.45 Overall place of release of consonant [k]

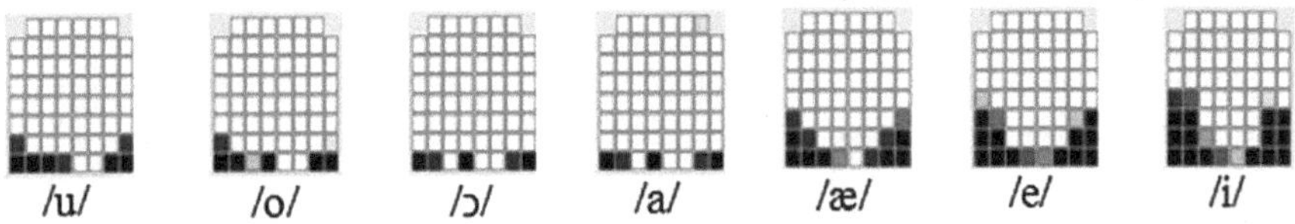

Fig. 2.46 Place of closure of /k/ for the Bangla vowels for the female speaker

made only up to the beginning of the soft palate. This problem will be seen for all manners for this place of articulation. However in the case of front vowels the complete closure is observed. Apparently the tongue hump is pulled forward for front vowels. Also there is a noticeable difference between the male and female informants, the tongue hump seem to be a little bit forwarded for the female speaker.

Figure 2.45 represents the overall place of release for the consonant /k/ with all the frames pooled together. Considering the limitation of the artificial palate mentioned above the figure indicates that the place of release may be taken as the traditional position of velar.

The area of contact of consonant /k/ varies with respect to different vowels. Average duration and energy in the occlusion is found to be 55.76 ms and −42.99 dB respectively. The spectral and signal domain evidences do not show voicing. This together with very low VOT (8.51 ms) makes this plosive unvoiced and unaspirated velar.

Figures 2.46 and 2.47 present EPG frames constructed in the manner described in an earlier paragraph depicting the place of closure of stop [k] for seven Bangla vowels respectively for the female and the male speakers. Here also is a noticeable

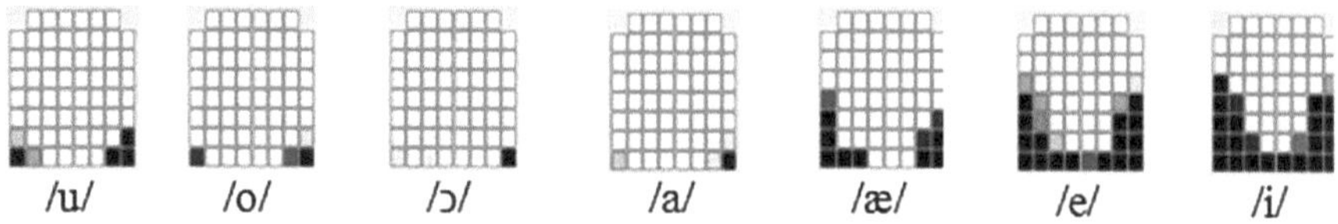

Fig. 2.47 Place of closure of /k/ for the Bangla vowels for the male speaker

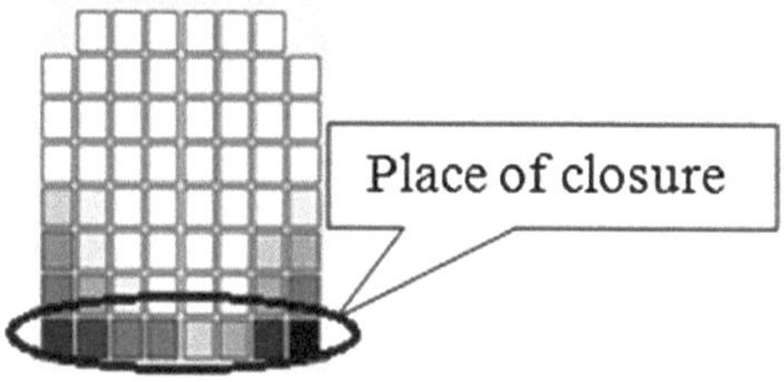

Fig. 2.48 Overall place of closure of consonant [k]

difference between the male and female informants, the tongue hump seem to be a little bit forwarded for the female speaker.

Figure 2.48 indicates that the overall place of closure for the stop [k]. Thus [k] is velar both as stop and plosive.

Henceforth the detail EPG frames for individual files shall not be shown in these sections. Only the overall frames computed from all frames separately for the male and female speakers will be presented in these sections. However for the interest of the inquisitive readers these will be provided with appropriate concise comments in the Appendix.

Plosive/Stop খ $[k^h]$:

Same difference as was seen for [k] between the male and female informants is present for $[k^h]$ also. Figures 2.49 and 2.50 represent the overall places of articulation respectively for release and closure for the plosive $[k^h]$. Both being almost identical $[k^h]$ is velar both as stop and plosive.

There is not much of a difference in closure configuration between [k] and $[k^h]$ except for a feeble indication that tongue may be a wee bit retracted for the aspirated counterpart.

The area of contact of consonant $[k^h]$ varies with respect to different vowels. Average occlusion duration is found to be 50.23 ms and average occlusion period

Fig. 2.49 Overall place of release of Consonant $/k^h/$

Fig. 2.50 Overall place of closure of consonant [k^h]

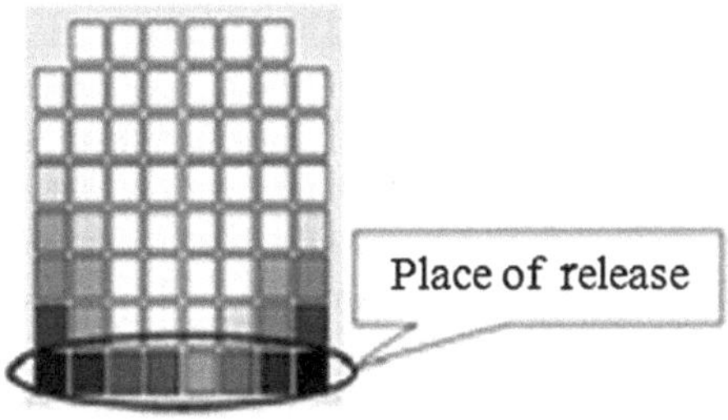

Fig. 2.51 Overall place of release of consonant [g]

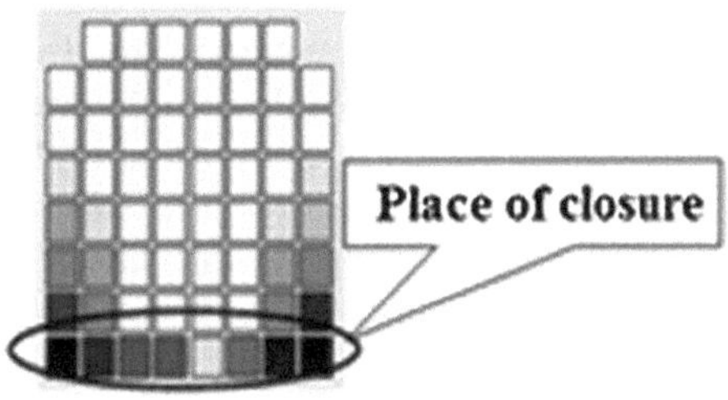

Fig. 2.52 Overall place of closure of consonant [g]

energy is −43.16 dB. The length of the VOT is 46.63 ms which is much larger than 8.51 ms of the unaspirated counterpart [k]. This makes this sound cognitively aspirated. There is no acoustic evidence of voicing during the occlusion. Thus both as a plosive and a stop Bangla [k^h] is unvoiced and aspirated.

Plosive/Stop গ [g]
Same difference as was seen for [k] between the male and female informants is present for [g] also. Figures 2.51 and 2.52 represent the overall places of articulation respectively for release and closure for the plosive [g]. Both being almost identical [g] is velar both as stop and plosive.

Average occlusion duration is found to be 44.42 ms and average occlusion period energy is −31.52 dB. In case of unvoiced counterpart the average energy was −42.99 dB. Thus occlusion region show significant acoustic energy. Also from the acoustic evidence it can be stated that during the occlusion period the glottal vibration do exist. There is also a negligible VOT of 4.47 ms after the burst along with the voiced signal. This duration is non-cognitive as far as aspiration is concerned. All these taken together shows that Bangla consonant [g] is voiced unaspirated velar.

Plosive/Stop ঘ [g^{h}]
Same difference as was seen for [k] between the male and female informants is present for [g^{h}] also. Figures 2.53 and 2.54 represent the overall places of articulation respectively for release and closure for the plosive [g^{h}]. Both being almost identical [g^{h}] is velar both as stop and plosive.

Average occlusion duration is 44.99 ms and average occlusion period energy is −33.01 dB. In case of unvoiced counterpart the average energy was −43.16 dB, which is much smaller. The acoustic evidence shows that during the occlusion period the glottal vibrations do exist. VOT after the burst being long (about 45.89 ms) aspiration is cognizable. Thus Bangla plosive [g^{h}] is voiced and aspirated.

In general for all manners the average EPG frames for velars, even when examined for each vowel, is almost the same (detail vowel-wise and sex-wise EPG frames are given in Appendix). The effect of the preceding vowels in the closure pattern is generally same, i.e. closure area is pulled forward slightly for front vowels. Similarly for all manners of production the place of articulation is slightly forwarded for the female speaker.

Plosive/Stop ট [ʈ] and ঠ [ʈ^{h}]
Figures 2.55 and 2.57 represents the overall place of release for the consonant [ʈ] and [ʈ^{h}] respectively. While the main concentration is in the Post alveolar region significant extension is observed up to the root of the teeth.

Figures 2.56 and 2.58 represents the overall place of closure for the consonant [ʈ] and [ʈ^{h}] respectively. A comparison of the above two sets of figures reveals noticeable difference. The place of closure for this consonant may still be

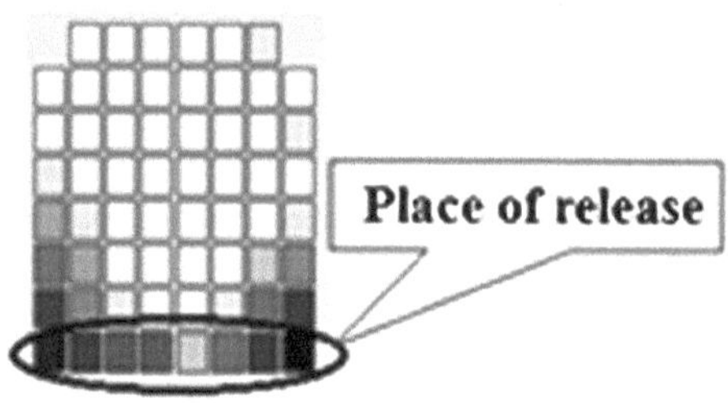

Fig. 2.53 Overall place of release of consonant[g^{h}]

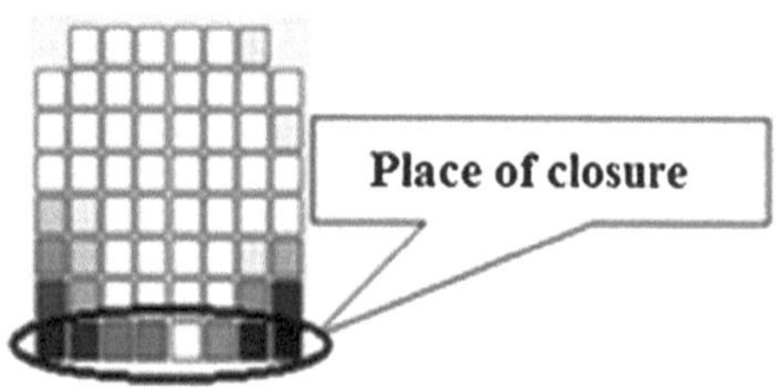

Fig. 2.54 Overall place of closure of consonant [g^{h}]

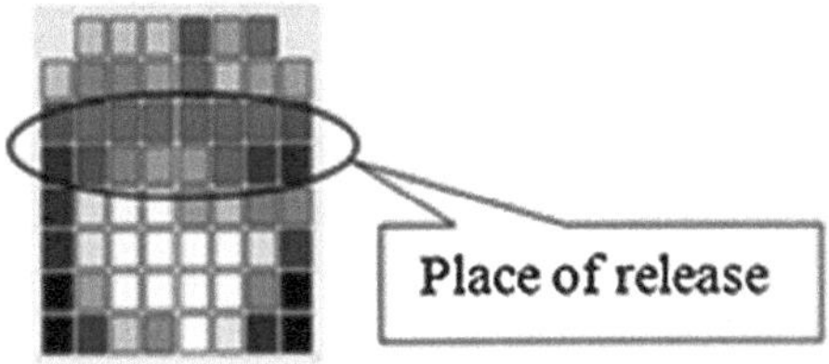

Fig. 2.55 Overall place of release of consonant [ʈ]

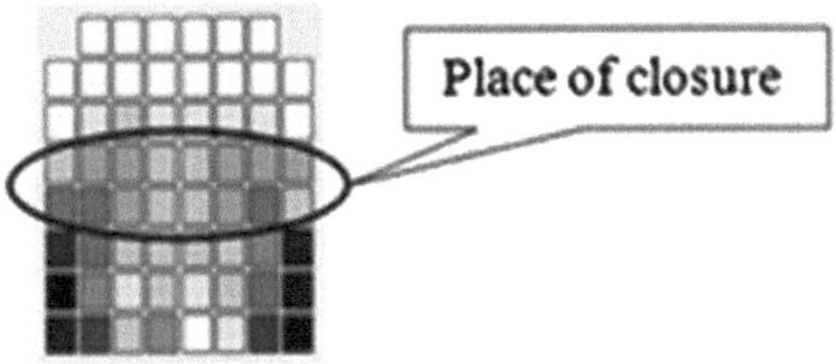

Fig. 2.56 Overall place of closure of consonant [ʈ]

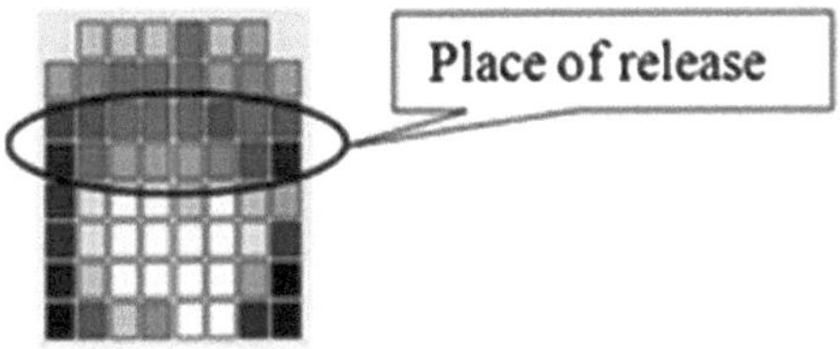

Fig. 2.57 Overall place of release of consonant [$ʈ^h$]

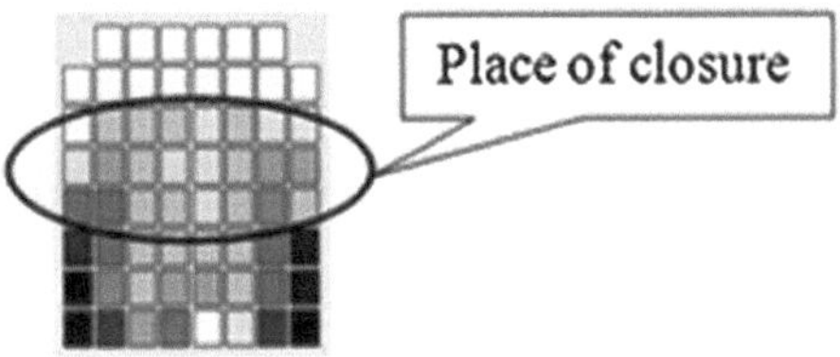

Fig. 2.58 Overall place of closure of consonant [$ʈ^h$]

considered as post-alveolar but distinctly palatalized. Thus for these plosives the tongue distinctly roles forward for release. Moreover one can notice that more area comes under contact in the case of release.

From the detailed vowel-wise EPGs (see Appendix) it is observed that the place of closure is post alveolar in case of front vowel of both speaker but it is palatal in case of back vowel for both the speaker.

Average occlusion duration for [ʈ]is 59.77 ms and average occlusion period energy is −41.91 dB. No acoustic evidence of voicing during occlusion is found also the length of the VOT (3.06 ms) being insignificant the manner of this plosive is unvoiced unaspirated.

Average occlusion duration for [$ʈ^h$] is 58.26 ms and average occlusion period energy is −43.22 dB. The significant length 44.05 ms of the VOT and absence of

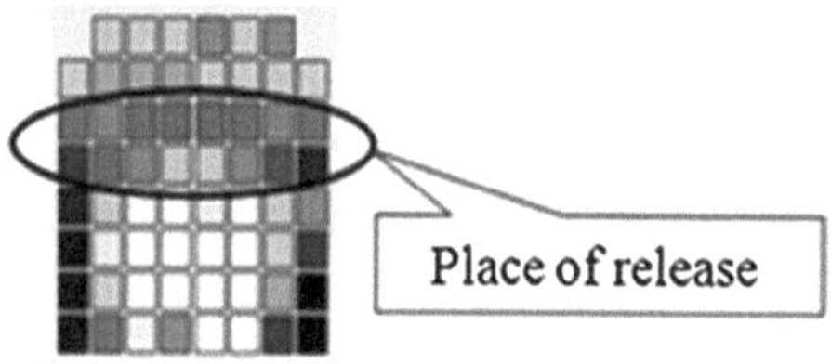

Fig. 2.59 Overall place of release of consonant [ɖ]

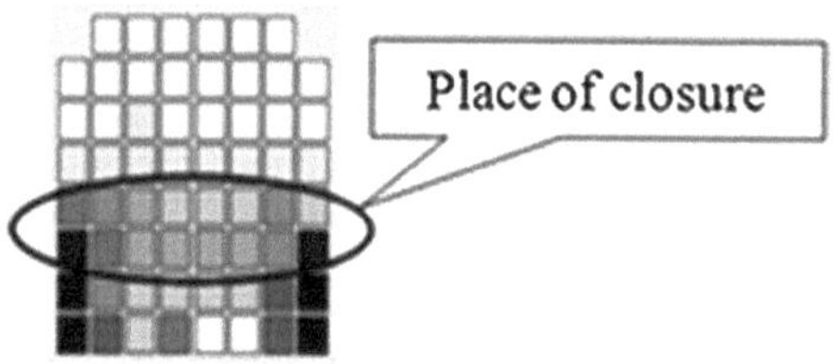

Fig. 2.60 Overall place of closure of consonant [ɖ]

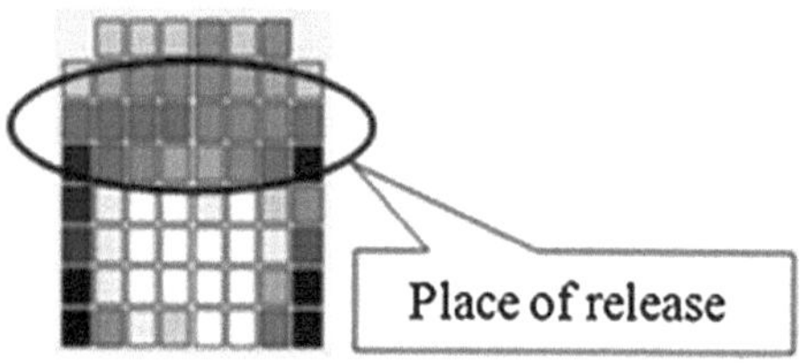

Fig. 2.61 Overall place of release of consonant [ɖ^h]

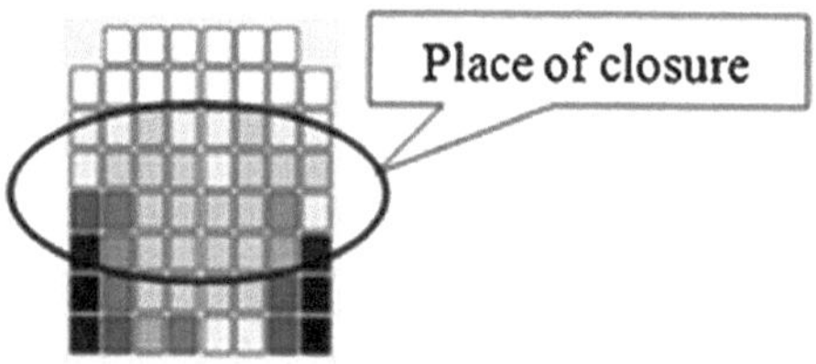

Fig. 2.62 Overall place of closure of consonant [ɖ^h]

the acoustic evidence of glottal vibration during the occlusion makes Bangla consonant [ʈ^h] as unvoiced aspirated post-alveolar as a plosive as well as a stop.

Plosive/Stop ড [ɖ] ঢ

Figures 2.59 and 2.61 represent the overall place of release for the consonant [ɖ] and [ɖ^h]. From this figure the place of these plosives is seen primarily to be post alveolar however one notices like in the case of its unvoiced counterpart some extension up to the root of teeth.

A perusal of Figs. 2.60 and 2.62 shows that the place of closure is palatal. From the detailed vowel-wise EPGs (see Appendix) it is observed that the place of closure is post alveolar in case of front vowel of both speakers but it is palatal in case of back vowel for both of them.

Average occlusion duration for [ɖ] is 59.57 ms and average occlusion period energy is −30.99 dB. From the acoustic evidence it can be stated that during the occlusion period the glottal vibration do exist. The average length of the VOT being 3.31 ms the plosive is unaspirated. So it can be concluded that the Bangla consonant [ɖ] is voiced unaspirated retroflex post-alveolar plosive.

Average occlusion duration of [ɖ^{h}] is 54.33 ms and average occlusion period energy is −28.67 dB. In case of unvoiced counterpart the average energy is −43.22 dB, which is much smaller. From the acoustic evidence it can be stated that during the occlusion period the glottal vibration do exist. There is also an aspiration of 48.7 ms after the burst along with the voiced signal. The manner of articulation is known to be retroflex. So it can be concluded that the Bangla consonant [ɖ^{h}] is voiced aspirated retroflex post-alveolar plosive/stop.

It is obvious that EPG cannot provide direct evidence of retroflexion. Therefore one has to depend on the subjective judgment of the speaker or an appropriate video-graphic arrangement for this. However if one carefully peruses and compares the overall EPG frames of all the plosive/stop consonant of Bangla one may notice significantly large increase in the area of contact for the retroflex ones, irrespective of the manners of production. This may be due probably to the fact that retroflection induces a larger and firmer contact with the palate. This then can be taken as the signature for retroflexion in EPG. The other signature appears to be that the release place is distinctly forwarded from that of closure indicating a forward movement of the tongue.

Plosive/Stop ত,থ,দ,ধ [t, t^{h}, d, d^{h}]
Figures 2.63, 2.64, 2.65, 2.66, 2.67, 2.68, 2.69 and 2.70 represent the overall place of release and closures for the consonant [t, t^{h}, d, d^{h}].

It may be observed from the above figures that the places of release of these plosives are at the region of root of teeth. Since the first sensor row of the artificial

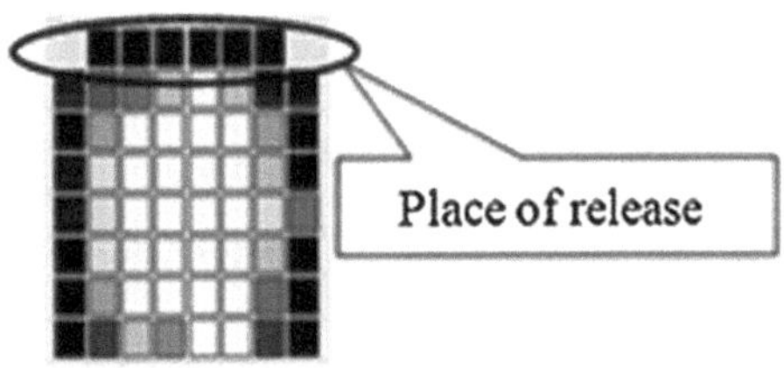

Fig. 2.63 Overall place of release of consonant [ʈ]

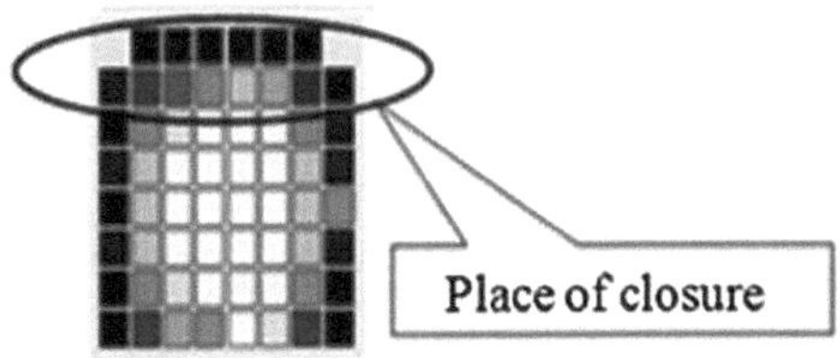

Fig. 2.64 Overall place of closure of consonant [ʈ]

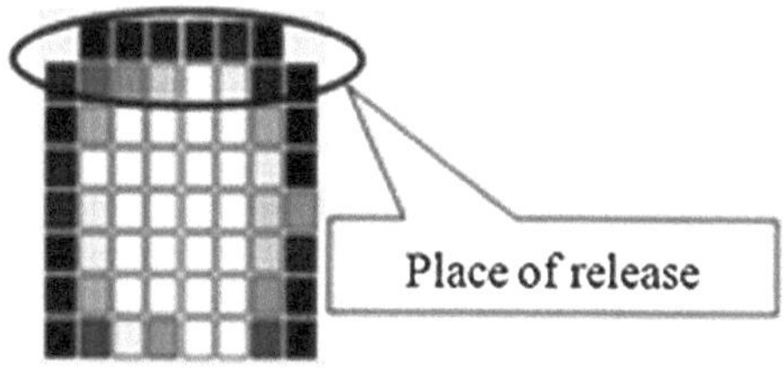

Fig. 2.65 Overall place of release of consonant $[ʈ^{h}]$

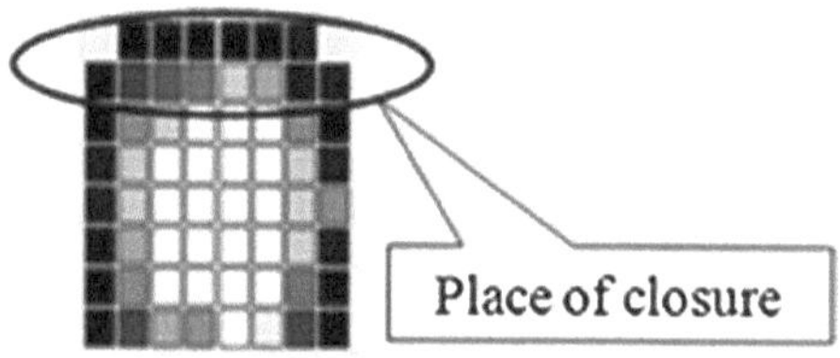

Fig. 2.66 Overall place of closure of consonant $[ʈ^{h}]$

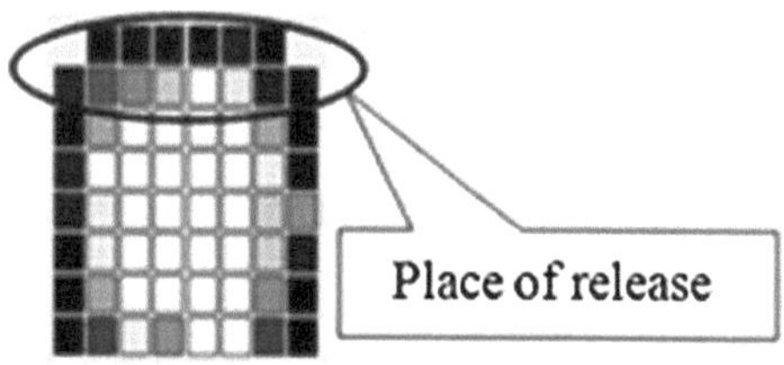

Fig. 2.67 Overall place of release of consonant [d]

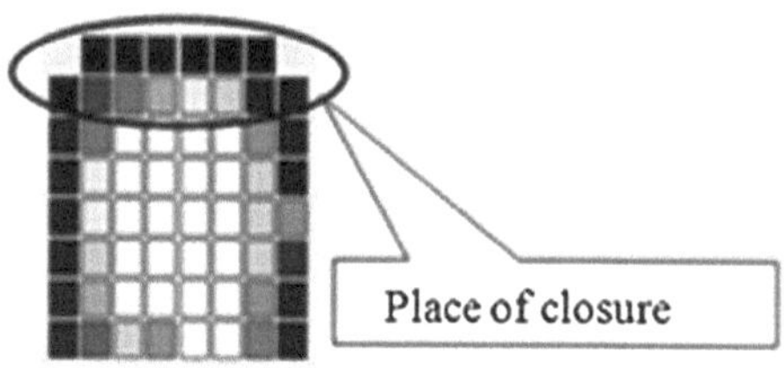

Fig. 2.68 Overall place of closure of consonant [d]

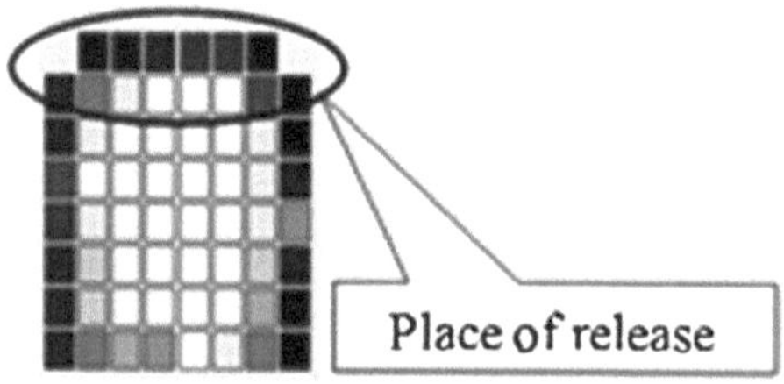

Fig. 2.69 Overall place of release of consonant $[d^{h}]$

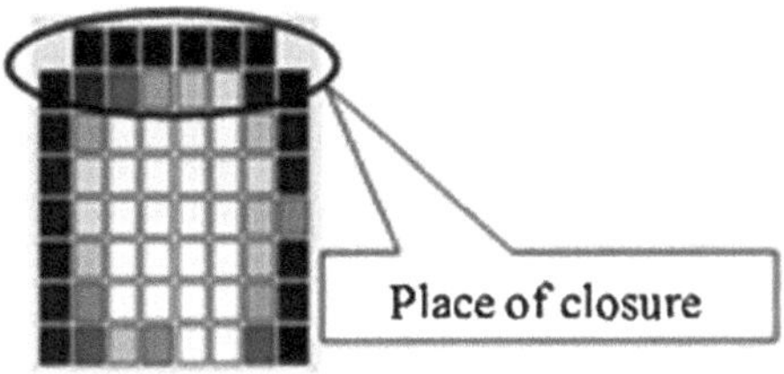

Fig. 2.70 Overall place of closure of consonant [d^h]

palate are only in the root of teeth region (there being no contact for teeth) and contact is always present in this region for all of the vowels this may possibly be considered as dental as it is traditionally believed.

From the above figures it is observed that the place of closure is somewhat retracted but still may be called dental.

Average occlusion duration is 64.93 ms and average occlusion period energy is −43.76 dB. There is no evidence of voicing. The length of the VOT is 3.94 ms which is negligible. Thus the Bangla consonant [t] is unvoiced unaspirated **dental** plosive/stop.

Average occlusion duration for [t^h] is 49.32 ms and average occlusion period energy is −42.56 dB. There is no acoustic evidence of voicing. The length of the VOT is 28.02 ms which makes aspiration cognizable. Thus the Bangla consonant [t^h] is unvoiced aspirated **dental** plosive.

Average occlusion duration for [d] is 48.86 ms and average occlusion period energy is −31.06 dB. From the acoustic evidence it can be stated that during the occlusion period the glottal vibration do exist. There is also an aspiration after the burst along with the voiced signal. The average length of the aspiration part is 3.03 ms and is insignificant. Thus the Bangla consonant [d] is voiced unaspirated dental plosive/stop.

Average occlusion duration for [d^h] is 42.05 ms and average occlusion period energy is −30.64 dB. From the acoustic evidence it can be stated that during the occlusion period the glottal vibration do exist. There is also 48.35 ms of VOT after the burst along with the voiced signal which makes aspiration cognizable. Thus the Bangla consonant [d^h] is voiced aspirated dental plosive.

The limitation in artificial palate to record properly the closure in the velar region has already been mentioned. One may also note here that because the artificial palate does not cover the teeth and only ends at the root of the teeth one is not sure whether some plosive/stops are purely dental.

The vowel-wise EPGs (presented in the Appendix) reveal that both the place of release and closure for both male and female speakers for front vowels the place of contact is forwarded. Though this is not large enough to merit particular attention in general, it is found to be significant in the following situation for retroflex plosives/stops:

(a) [ʈ, ʈh, and ɖ] are dental with male speaker for release in front vowels
(b) [ʈ, and ʈh] are palatal/mediopalatal with both speakers in closure of back vowels

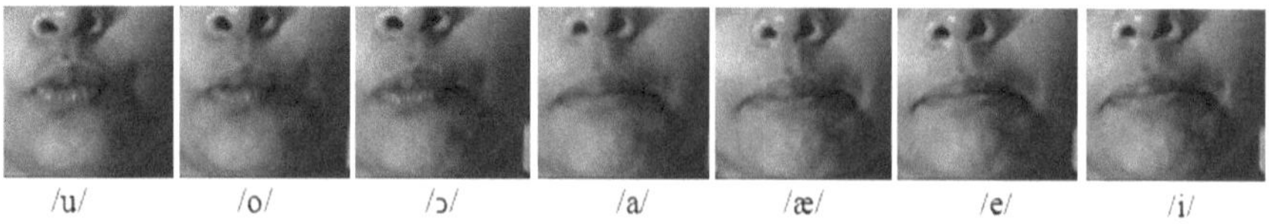

Fig. 2.71 The release frames of lips for [p]

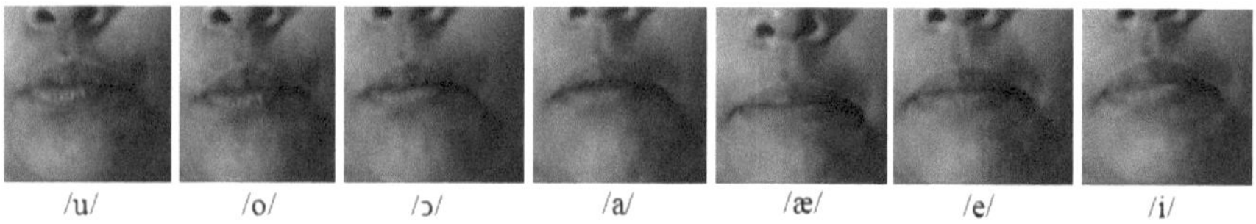

Fig. 2.72 The release frames of lips for [p^h]

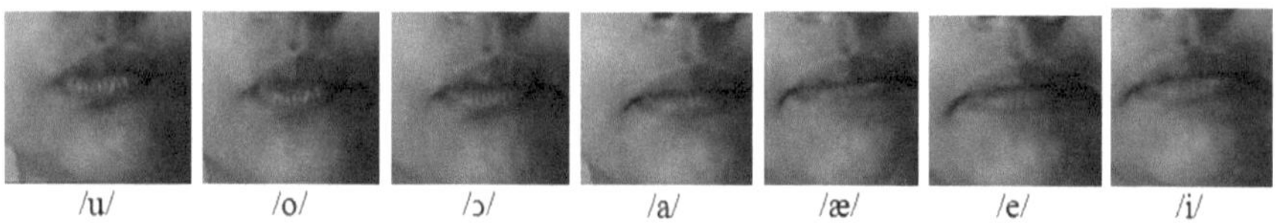

Fig. 2.73 The release frames of lips for [b]

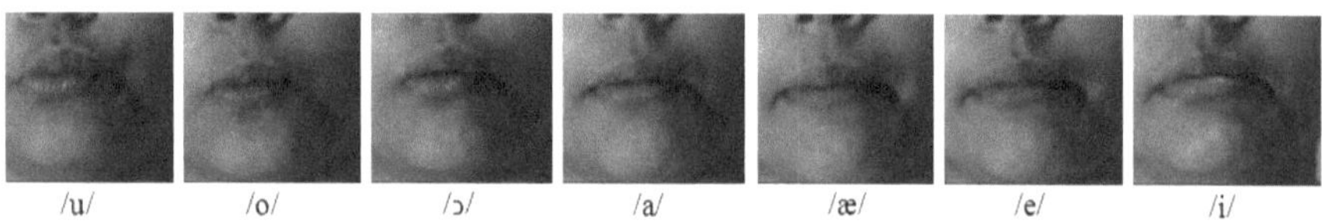

Fig. 2.74 The release frames of lips for [b^h]

(c) [d̪h] is palatal/mediopalatal with male speaker in closure of back vowels

Plosive/Stop প, ফ, ব, ভ [p, p^h, b, b^h]
Traditionally [p, p^h, b and b^h] are known as bilabials and as expected the EPG did not show any closure in the entire artificial palate region. The video of lips movements of the informant is recorded during the production of this sound and from the video evidence both the release and closure are found in the lips region. Figures 2.71, 2.72, 2.73 and 2.74 represent the lips release frame for these plosives. These indicate that these plosives are bilabial.

The varied degree of rounding of the lips corresponding to the vowels may be observed from the squeezing of the lips in the pictures.

Table 2.7 Summary of occlusion parameters VOT

	Occlusion		VOT(ms)		Occlusion		VOT(ms)
	Time (ms)	Energy(dB)			Time(ms)	Energy(dB)	
k	55.76	−42.99	8.51	g	44.42	−31.52	4.47
kh	50.23	−43.41	46.63	gh	44.99	−33.01	45.89
ʈ	59.77	−41.91	3.06	ɖ	59.57	−30.99	3.31
ʈh	58.26	−43.22	44.05	ɖh	54.33	−28.67	48.7
t	64.93	−43.76	3.94	d	48.86	−31.06	3.03
th	49.23	−42.56	28.02	dh	42.05	−30.64	48.35
p	70.97	−43.32	3.58	b	51.72	−30.64	2.76
ph	47.51	−41.23	32.11	bh	49.59	−29.94	52.51
Average	57.08	−42.8	NA		49.44	−30.81	NA

Average occlusion duration for [p] is 70.49 ms and average occlusion period energy is −43.32 dB. No acoustic evidence of voicing was found. The length of the VOT, 3.58 ms, is insignificant. Thus Bangla [p] is unvoiced unaspirated bilabial.

Average occlusion duration for [p^h] is 47.41 ms and average occlusion period energy is −41.23 dB. The length of the VOT is 32.11 ms which indicates cognitive level of aspiration. There is no acoustic evidence of glottal vibration during the occlusion period indicating this Bangla consonant to be unvoiced aspirated bilabial plosive.

Average occlusion duration for [b] is 51.72 ms and average occlusion period energy is −30.64 dB. There is also an aspiration after the burst along with the voiced signal. However the VOT of 2.76 ms is insignificant. Along with these, glottal vibration during the occlusion period indicate that the Bangla consonant [b] is voiced unaspirated bilabial.

Average occlusion duration for [b^h] is 49.59 ms and average occlusion period energy is −29.94 dB. The average length of the VOT is 52.51 ms is significant. Along with this glottal vibration during the occlusion period indicate that the Bangla consonant [b^h] is voiced aspirated bilabial.

Table 2.7 gives the summary of parameter values for the occlusion period and VOT for all plosives. The occlusion time is 14% less for voiced counterpart for all articulatory positions. This difference is known to be cognizably significant. The acoustic energy during occlusion is significantly less (12 dB approximately) for unvoiced counterpart. VOT is negligible for unaspirated counterpart and cognizably significant (except for [t^h] and [p^h]. These figures show that manner based classification of Bangla plosives are phonetically significant.

2.5.2 *Fricative*

The place of articulation of the friction can be determined from the EPG frame where the opening is narrowest (see Fig. 2.41). The method of selection of appropriate frame is described earlier in Sect. 1.5.

Fricative শ [ʃ]
Figure 2.75 presents the overall place of constriction for the fricative [ʃ].

It may be seen from the above figure that the overall place of constriction for the consonant [ʃ] is post-alveolar. There being no acoustic evidence of voicing it may be said that /ʃ/ is Unvoiced Post alveolar Fricative

There has been no noticeable overall significant difference between male and female informants. However it has been noticed that in case of male speaker the place of constriction is slightly palatalized for back vowels.

Fricative স [s]
Figure 2.76 presents the overall place of constriction for the fricative [s].

The place of constriction of the consonant [s] is primarily alveolar though extends up forward a little. There being no acoustic evidence of voicing this consonant is unvoiced alveolar fricative. For the informants used the constriction appears to be on the right side of the mouth.

Examination of detail EPGs reveal that in case of front vowel the constriction area is little bit forwarded. In the case of male speaker the constriction area spread up to the root of the teeth region.

Fricative ষ [ʂ]
Figure 2.77 represents the overall place of friction for the consonant [ʂ] which is palatal. The retroflexion is not revealed in EPG; however this fricative is traditionally believed to be retroflexed.

Here being no acoustical evidence of voicing [ʂ] may be regarded as unvoiced retroflexed palatal fricative.

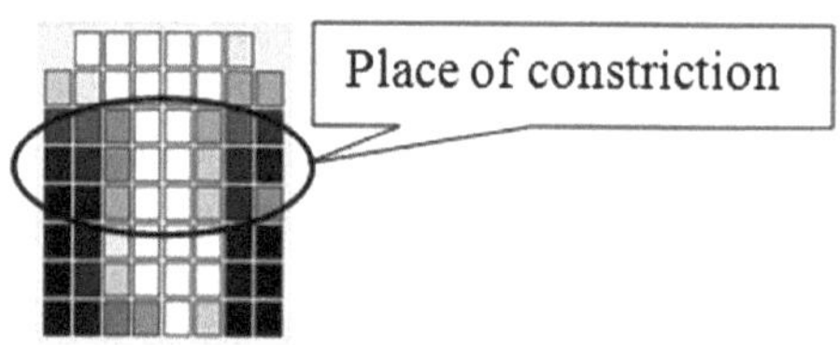

Fig. 2.75 Overall place of constriction of consonant /ʃ/

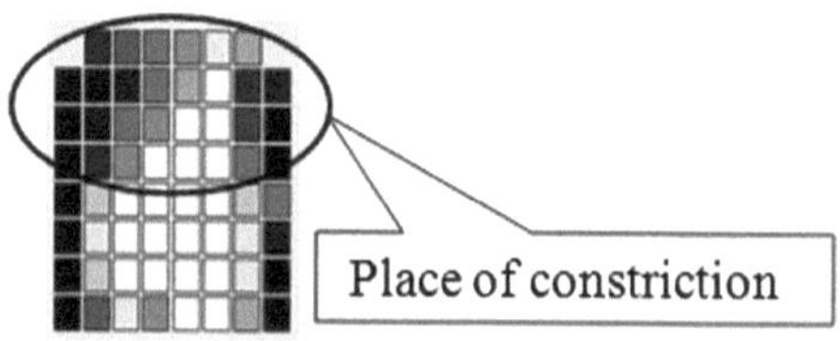

Fig. 2.76 Overall place of constriction of consonant [s]

Fig. 2.77 Place of constriction of [ʂ] for pooled data

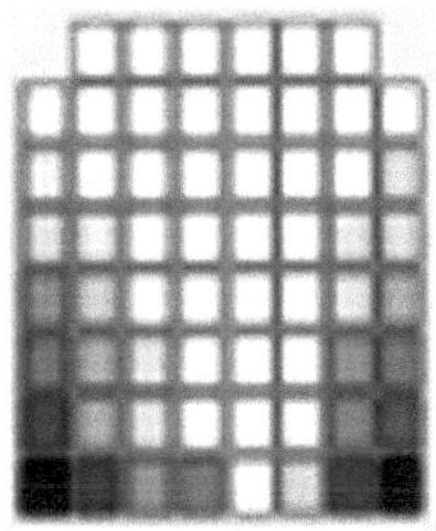

Fig. 2.78 Overall EPG frame for [h]

There is a consistence EPG evidence of the place of articulation of all fricatives in conjunction with front vowels, particularly for the male speaker, to be forwarded.

Fricative হ [h]
Figure 2.78 represents over all EPG frame contact position for all seven aspirated vowels due to the constriction in association with the phone [h].

From the above figures it may be observed that there is an indication of constriction appearing at the lower end of artificial palate. This strongly suggests a constriction below the velar region. The traditional literatures also present [h] as a glottal fricative.

However a recent study of this sound in Bangla (Datta 2014) revealed that in this dialect [h] does not reveal itself as a fricative but causes the succeeding vowel murmured (see Sect. 1.5 of Chap. 1). In fact due to its influence a set of seven murmured vowels has been reported for Bangla.

2.5.3 *Affricates*

In the case of affricates the place of release of the occlusion is determined from the EPG frame just before the opening of the closure. The place of articulation of the friction is determined by the EPG frame where the opening is narrowest. Manner of articulation of the affricates is determined from the acoustic study of the above-mentioned segments. The average occlusion duration and VOT duration along with the average energy at occlusion period is presented.

Affricate চ[ʧ]
An affricate is a complex sound in the sense that it contains two consecutive articulatory events namely plosion and friction, physically a sharp short release

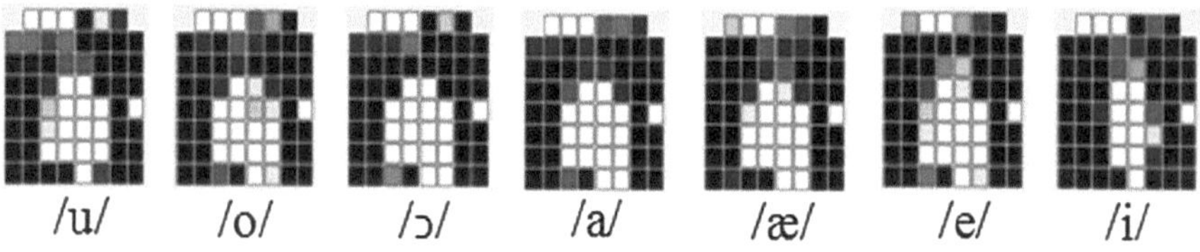

Fig. 2.79 Place of release of [ʧ] for seven Bangla vowels of female speaker

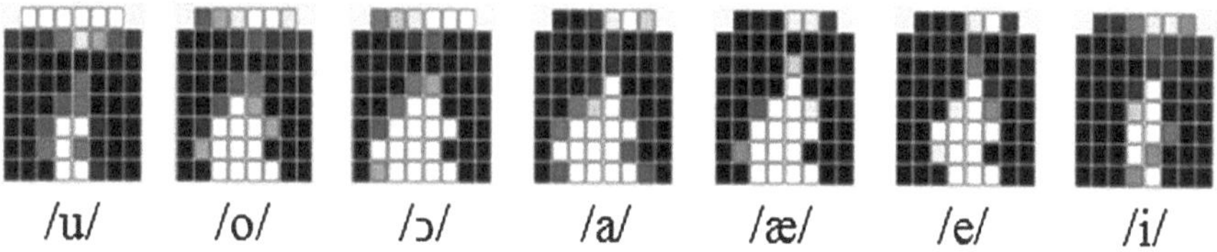

Fig. 2.80 Place of release of [ʧ] for seven Bangla vowels of male speaker

Fig. 2.81 Overall place of release of [ʧ]

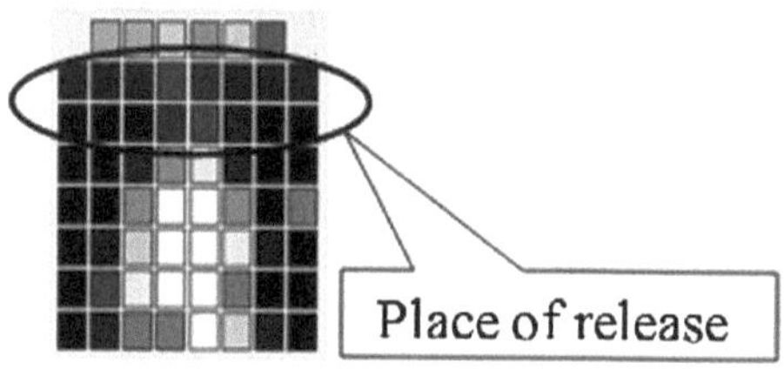

followed by a constriction before fully opening for the vowel. During the whole period the tongue may roll. Thus there may be two different places where tongue is engaged with the palate. This calls for citing both of them as places of articulation. However when this consonant is considered as the stop one may take the place of closure as the place of articulation. Similarly, when this is at the beginning of the vowel the place of constriction may be taken as the place of articulation. This convention is followed here.

Figures 2.79 and 2.80 represents the place of release of [ʧ] for seven Bangla vowels spoken respectively by the female and male speakers. Though there is no mentionable difference in the place of articulation the male speaker show greater amount tongue-palate contact.

Figure 2.81 represents the overall place of release.

From the above figures it is observed that the place of release is alveolar for the back vowel in both male and female speaker. In the case of front vowels the tongue tip is little bit forwarded during the release.

Figures 2.82 and 2.83 represents the place of constriction, for generating friction, in [ʧ] for seven Bangla vowels spoken respectively by the female and male speakers.

Figure 2.84 represents the overall place of constriction.

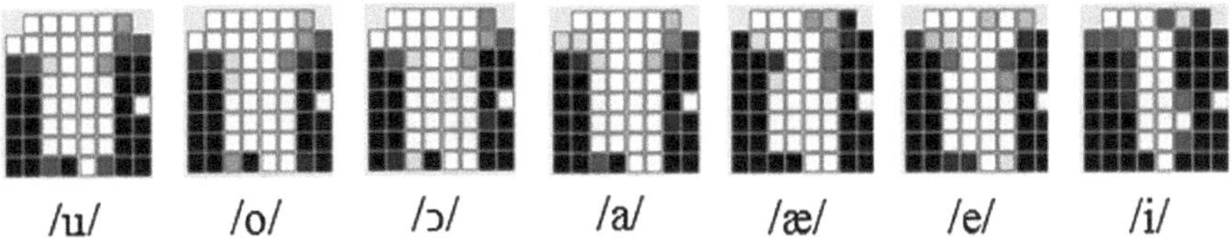

Fig. 2.82 Place of constriction of [ʧ] for seven Bangla vowels of female speaker

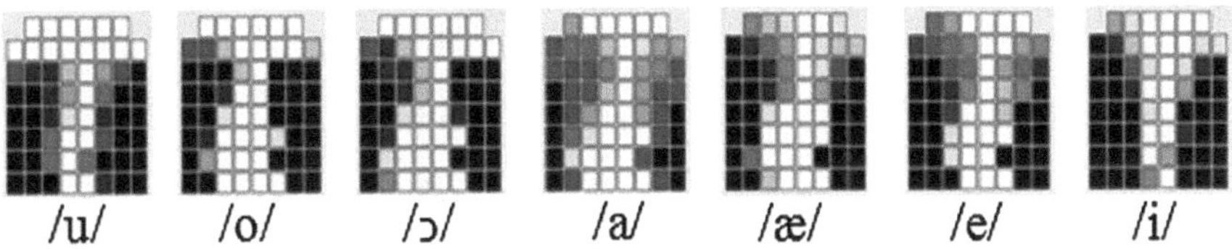

Fig. 2.83 Place of constriction of [ʧ] for seven Bangla vowels of male speaker

Fig. 2.84 Overall place of constriction of [ʧ]

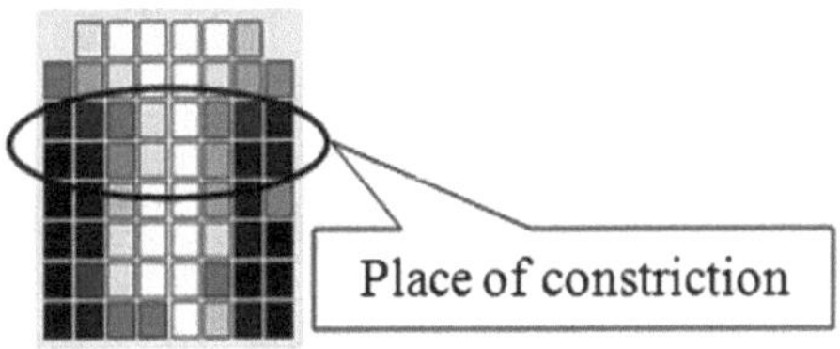

Though there is no mentionable sex-wise difference in the place of articulation the male speaker show greater amount tongue-palate contact. The place of constriction is post alveolar.

Thus for this affricate occlusion occurs at alveolar position and constriction occurs at post alveolar position.

Average occlusion duration is 41.98 ms and average occlusion period energy is −42.51 dB. The length of the VOT (3.66 ms) is insignificant. Also there is no acoustic evidence of voicing. So it can be concluded that the Bangla consonant [ʧ] is unvoiced unaspirated affricate where the place of occlusion is alveolar and place of friction is post-alveolar.

Affricate ছ [ʧ^{h}]

Figures 2.85 and 2.86 represents the overall place respectively of release and of the friction for the fricative [ʧ^{h}]. The place of release is seen as alveolar and that of friction is post-alveolar.

Figure 2.86 represents the overall place of friction of the fricative [ʧ^{h}], which is post-alveolar.

Average occlusion duration is 35.85 ms and average occlusion period energy is −42.94 dB. The length of the VOT is 27.23 ms large enough to be cognitively significant. There being no evidence of the glottal vibration it can be concluded that

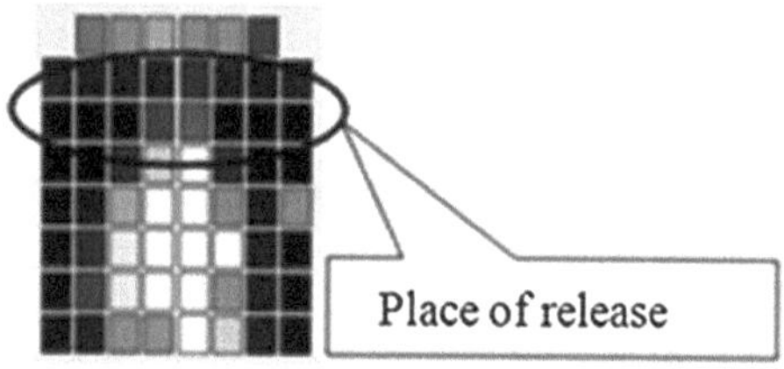

Fig. 2.85 The overall place of release of [ʧʰ]

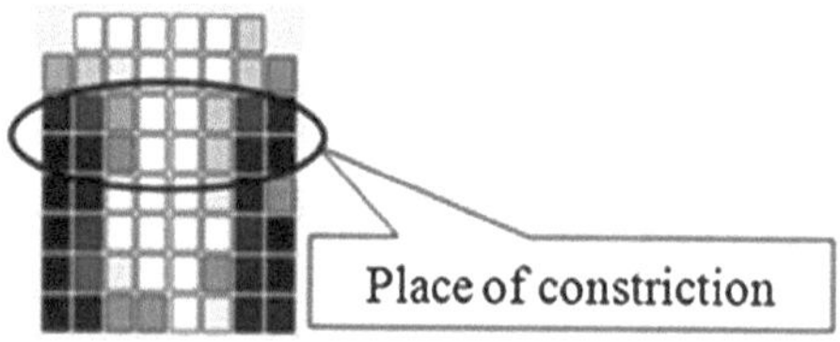

Fig. 2.86 The overall place of friction of [ʧʰ]

this Bangla consonant [ʧʰ] is unvoiced aspirated affricate where the place of occlusion is alveolar and place of friction is post-alveolar.

Affricate জ [ʤ]

Figures 2.87 and 2.88 represent the overall places respectively of release and constriction for the fricative [ʤ]. The place of release and friction is seen to be respectively at alveolar and post-alveolar region.

Average occlusion duration is 41.63 ms and average occlusion period energy is −30.07 dB. The average length of VOT (5.08 ms) is insignificant. This with the glottal vibration during the occlusion period indicates that the Bangla consonant [ʤ] is voiced unaspirated affricate where the place of articulation of occlusion is alveolar and place of articulation of friction is post-alveolar.

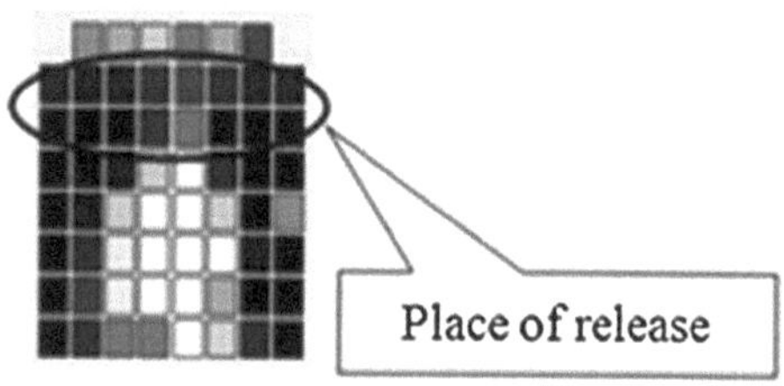

Fig. 2.87 Overall place of release [ʤ]

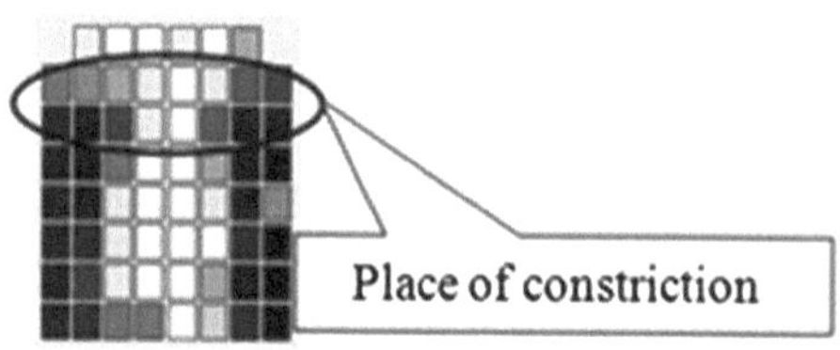

Fig. 2.88 Overall place of constriction [ʤ]

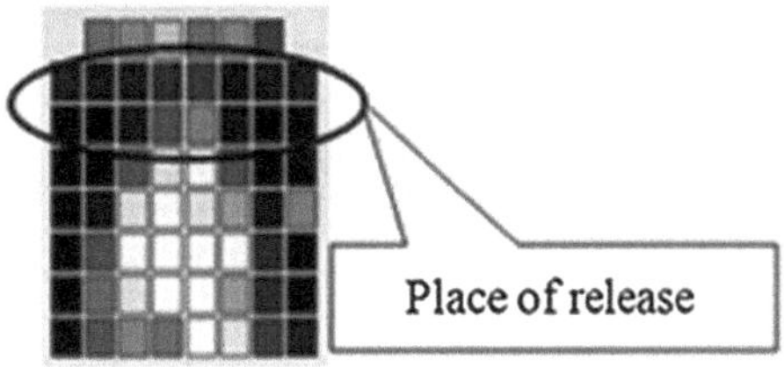

Fig. 2.89 Overall place of release of [ʤʰ]

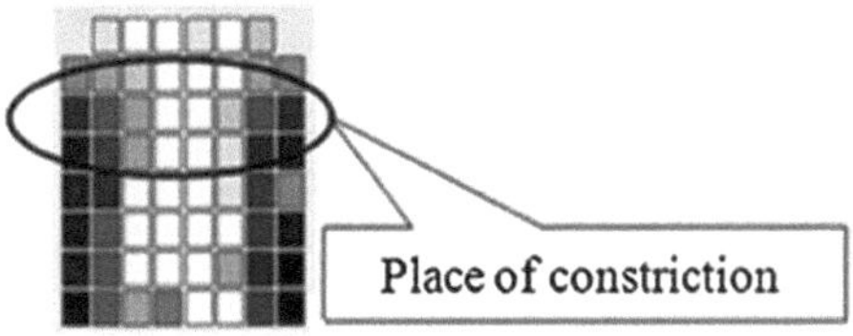

Fig. 2.90 Overall place of constriction of [ʤʰ]

Affricate ঝ [ʤʰ]
Figures 2.89 and 2.90 represents the overall place of release of [ʤʰ] and place of friction respectively

As the place of release and place of friction together represents the place of articulation then the place of articulation of release is alveolar and place of articulation of friction is post alveolar. Average occlusion duration is 34.04 ms and average occlusion period energy is −32.51 dB. From the acoustic evidence it can be stated that during the occlusion period the glottal vibration do exist. There is also an aspiration after the burst along with the voiced signal. The average length of the aspiration part is 26.02 ms. Thus the Bangla consonant [ʤʰ] is voiced aspirated affricate where the place of articulation of occlusion is alveolar and place of articulation of friction is post-alveolar. From detail vowel-wise EPGs for Bangla affricates it is noticed that the tongue tip is little bit forwarded during the release for front vowels.

Thus Bangla affricates, irrespective of their manner of production are alveolar as stop and post alveolar as release. This indicates that during the production of these sounds the tongue is retracted by one place of articulation.

2.5.4 *Laterals*

The place of articulation of the lateral is determined by the EPG frame collected from the middle of the production of the consonant.

Lateral ল [l]
Figures 2.91 and 2.92 represent the EPG frames collected at middle of the production of the consonant [l] with all the seven vowels spoken by the female and male speakers respectively. Let us peruse the EPG frame corresponding to the

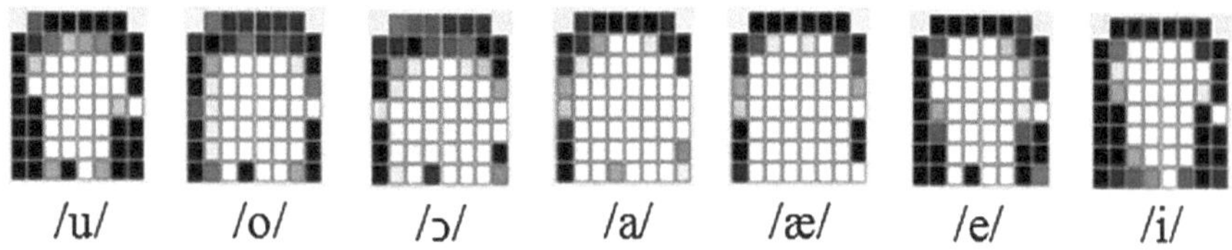

Fig. 2.91 EPG frame for all the seven vowel of female speaker

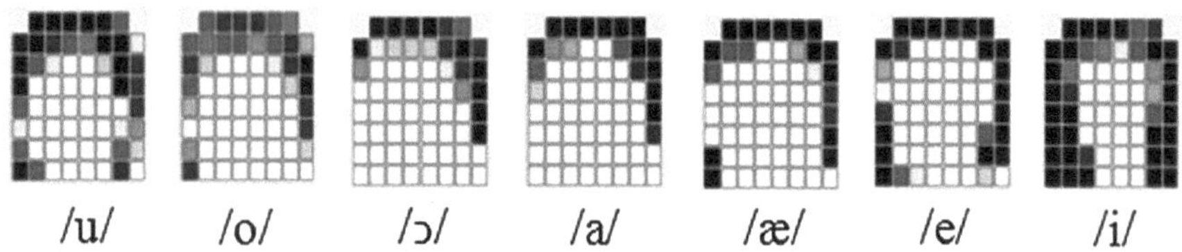

Fig. 2.92 EPG frame for all the seven vowels of male speaker

Fig. 2.93 Overall articulatory contact positions

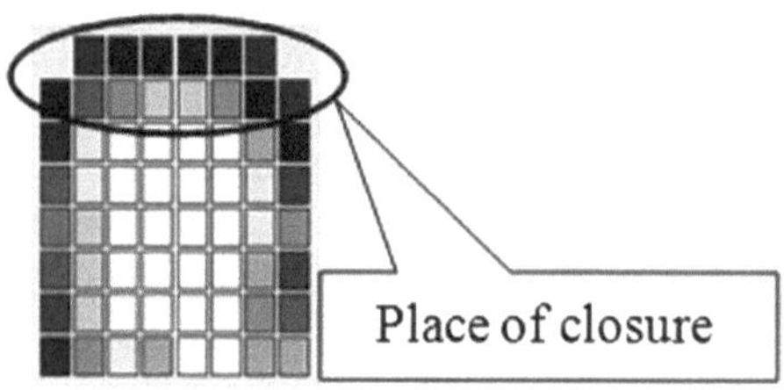

vowel /a/, one may note here that it is the gray level representation of the ten repetitions. There is a clear lateral path on the right flank for all utterances as indicated by the three white cells on the right. At the same times there are two gray cells on the left flank which indicates that for some utterance this pronunciation is bilateral. Thus from minute examination of the average frames one can gather the nature and the side of lateralization for the laterals.

Figure 2.93 represents overall articulatory contact positions for the consonant [l].

From the above figures it is observed that during the production of consonant [l] the articulator completely closes the air passage medially at front and air passes thorough the side of the tongue. If the air passes through the one side of the tongue then it is called unilateral and if the air passes through the both side then it is called bi-lateral. From the above figures one could say that the Bangla consonant [l] is predominantly unilateral with the caution that only two informants have been examined. The place of articulation is dental. The average duration of the [l] in Bangla is 54.02 ms.

2.5.5 Nasal Murmur

The place of articulation of the murmur is determined by the EPG frame collected from the middle of the production of the consonant.

Consonant ঙ [ŋ]
Figure 2.94 represents overall articulatory closure positions for the consonant /ŋ/.

From the detailed vowel wise figure given in the Appendix it is seen that a complete closure is observed in case of front vowels. Due to the limitation of artificial palate sensor position as mention in Sect. 2.5.1 the complete closure is not observe in case of rest of the vowels. From Fig. 2.99 the place of closure is in the velar region. Therefore the place of articulation of [ŋ] is velar. The average duration of the [ŋ] in Bangla is 50.68 ms

Consonant ঞ [ɲ]
The occurrence of this consonant is rare. Only two Bangla words (মিঞা, যাচ্ঞা) has been used for the study of this consonant. The 5 repetitions of those words are recorded and the EPG frame collected from the middle of the production of the consonant is considered for the place of articulation determination.

Figure 2.95 represents the overall place of closure for the above consonant.

From the figure it is observed that the place of articulation of the above consonant is alveolar.

Consonant ণ [ɳ]
In general the [ɳ] is not pronounced in isolation in Bangla. It is only pronounced in cluster with retroflex oral sounds. Due to the above reason this consonant is recorded within a word where it is pronounced instead of usual VCV combination.

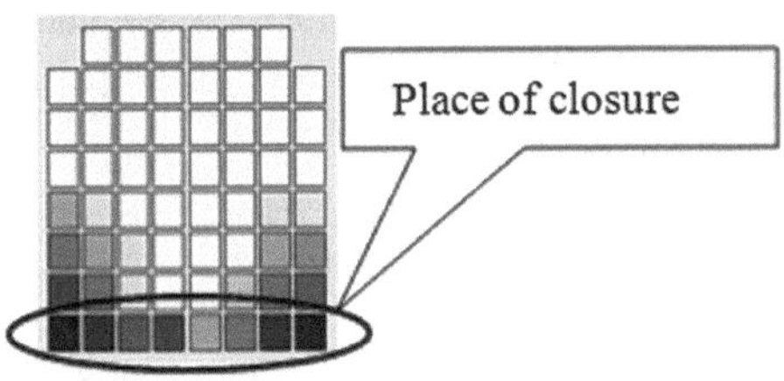

Fig. 2.94 Overall articulatory closure positions

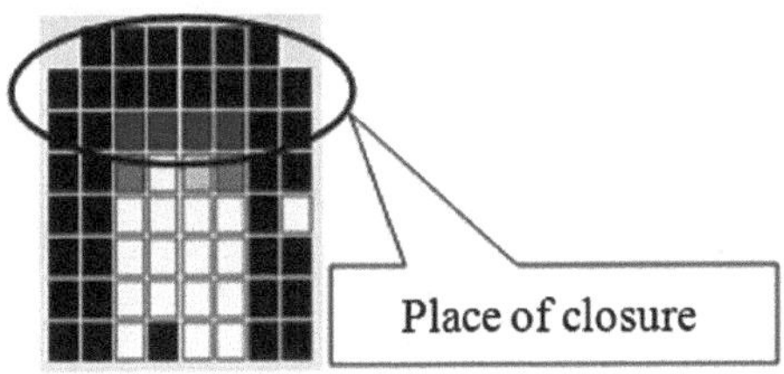

Fig. 2.95 Overall place of closure for /ɲ/

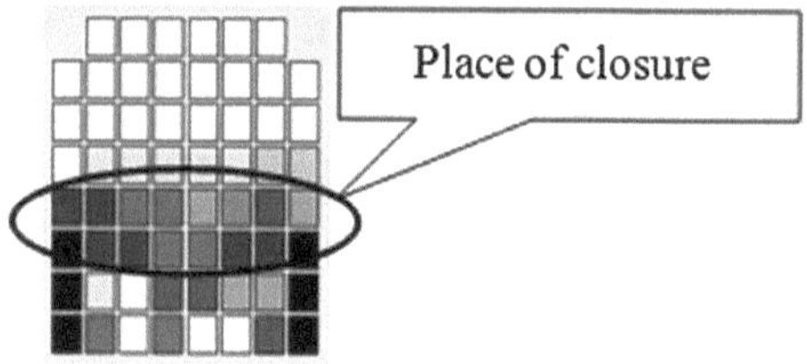

Fig. 2.96 Overall articulatory closure positions

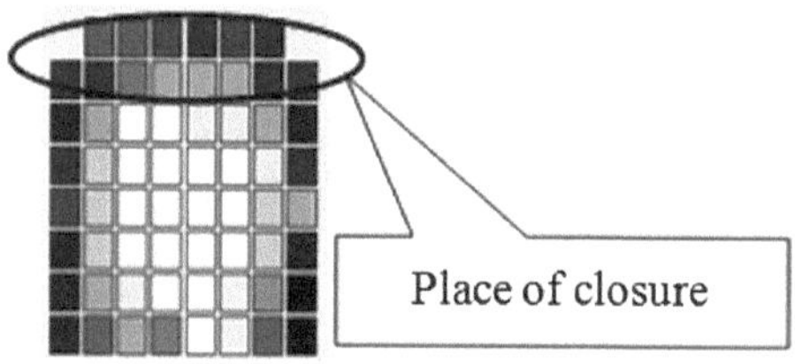

Fig. 2.97 Overall articulatory closure positions

EPG frames collected from the middle of the production of the consonant are considered for the determination of the place of articulation.

Figure 2.96 represents the overall EPG closure position for the above consonant.

From the above figure it is observed that the above consonant place of articulation is palatal. Its manner of production is retroflex nasal. The average contact area is 32.57%. The average duration of the [ɳ] in Bangla is 44.12 ms.

Consonant ন [n]

Figure 2.97 represents overall articulatory closure positions for the consonant [n].

The above figure shows that the place of contact this consonant is at the root of the teeth. Taking note of the deficiency of the palate in not having contacts on the teeth area, one may consider that the place of articulation of the above consonant as **dental**. The average duration of the [n] in Bangla is 54.65 ms.

Consonant ম [m]

After analyzing the EPG frame data it is observed that there is no closure in the whole palate region. The video of lips movement of the informant is recorded during the production of this sound and from the video evidence the closure is found in the lips region. Figure 2.98 represents the lips closure of during the production of the above sound for all the Bangla vowels. The figures show that the consonant [m] is bilabial. The average duration of the /m/ in Bangla is 57.06 ms

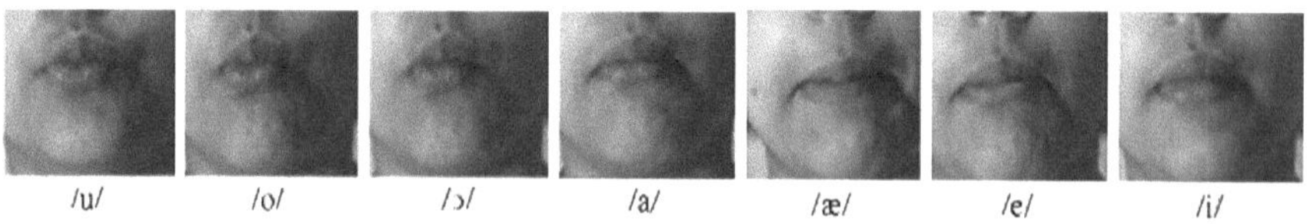

Fig. 2.98 Lips closure of during the production

In this experimental setup nasal spirometers to measure air passing out of nose was not available. Therefore the nasality of this group of consonants was determined from the acoustic evidence and subjective listening.

2.5.6 *Trills, Flap or Tap*

Consonant র [r]

Figure 2.99 represents overall articulatory positions for the consonant [r].

The detail vowel wise frames show that there is no complete closure region. This may be due to the situation that [r] being found mostly a tap (see Sect. 2.5.6) the momentary contact of the tongue tip may occasionally miss capture in the EPG system. Therefore the continuous black line indicating complete closure is not always visible. However the gray level connectivity reflects the occasional capture of the blockage. Overall place of closure can be considered at alveolar region. So the places of articulation of the Bangla trill [r] is **alveolar**.

From an examination of the detailed vowel-wise EPG it is observed that the place of contact of the tip of the tongue is more forwarded in case of front vowel for both the speaker.

Consonant ড় ***([ɽ])***

The place of articulation of these sounds is determined by the EPG frame collected from the middle of the production of the contact area. Contact area is measured by the number of contacts made to the total number of electrodes in the whole EPG frame and is expressed as percentage.

Figure 2.100 represents overall articulatory positions for the consonant [ɽ].

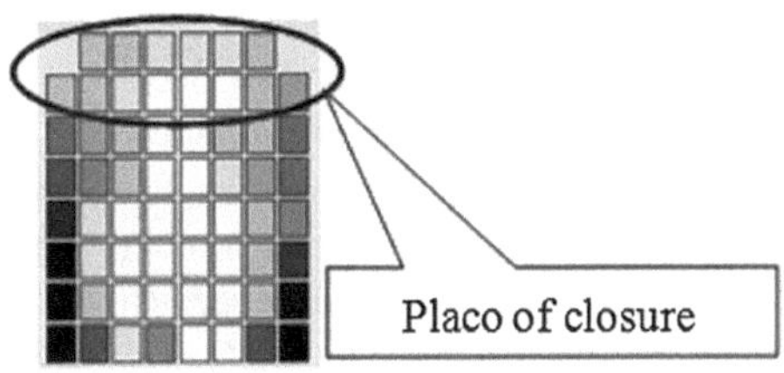

Fig. 2.99 Overall articulatory positions

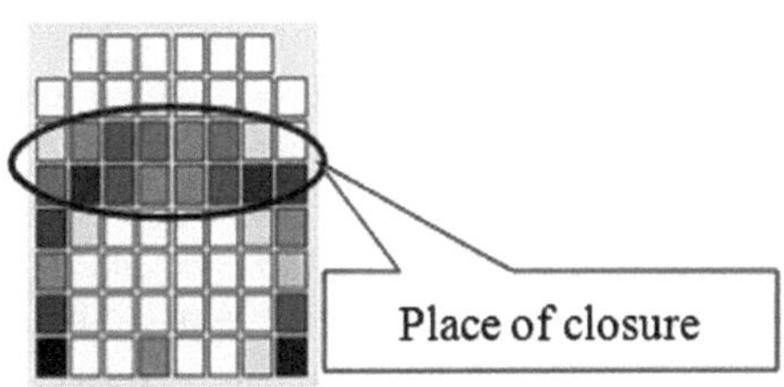

Fig. 2.100 Overall articulatory position of ([ɽ])

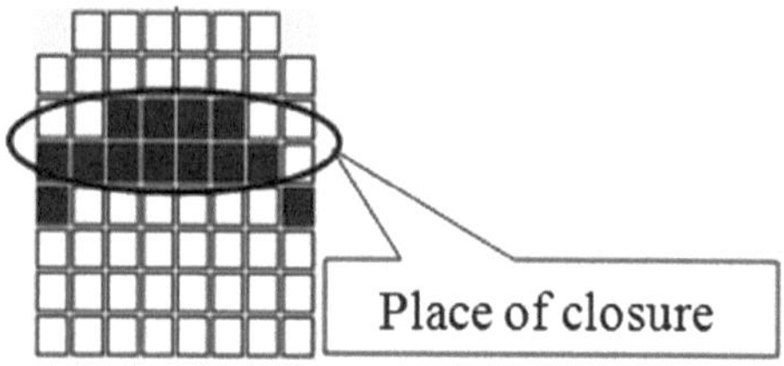

Fig. 2.101 Overall articulatory position of [ɽ^{h}]

The closure is in the region of post alveolar. The above consonant is retroflex flap and the place of articulation is **Post alveolar**.

Consonant ढ़ ***([ɽh])***
The consonant [ɽh] is the aspirated counterpart of the consonant [ɽ], the place of articulation remaining the same as for [ɽ]. But the only difference is that after the release there is a perceptible amount of aspiration. Figure 2.101 represents the overall EPG data for the contact position for [ɽ]. Thus [ɽh] is post alveolar aspirated retroflex tap.

The EPG analysis confirms the traditionally known places of articulation for [r, ɽ and ɽh]. However both EPG and acoustic analysis revealed the possibility of [r] being a tap and [ɽh] being a trill, which is a deviation from the commonly held belief.

2.5.7 Area of Obstruction

The consonants dealt with above are some form of obstruent where sometimes firm and complete obstruction takes place and other times narrow constriction are made which produced quasi-random signals. The role of area where obstruction of the passage occurs needs some understanding regarding the role of articulators in the production. It has been noticed from detailed vowel-wise EPGs that the position as well as the contact area is influenced by the adjoining vowels. The two important measures are area of actual contact as well as area of constriction. Figure 2.102 presents the average contact areas of appropriate types for different consonants. Velar stops are not included because the artificial palates do not cover the appropriate region. Similarly bilabials are not included because of obvious reasons. Taps and trills do not exhibit firm contacts.

A perusal of the figures reveals that the area of contact/constriction depends consistently on the adjoining vowel. The area is least for central low vowel [a]. The area rises from low to high. For low vowels the vertical movement is larger and therefore one may expect lesser time for the contact with the palate. This could be a possible explanation. Further there is a general trend of increase in the area from back to front, very sharply towards front. This is expected as with the front vowels the contact for the consonant is made by raising the tip of the tongue while for the vowel the portion near the root is raised. This indicates a complex motion for the

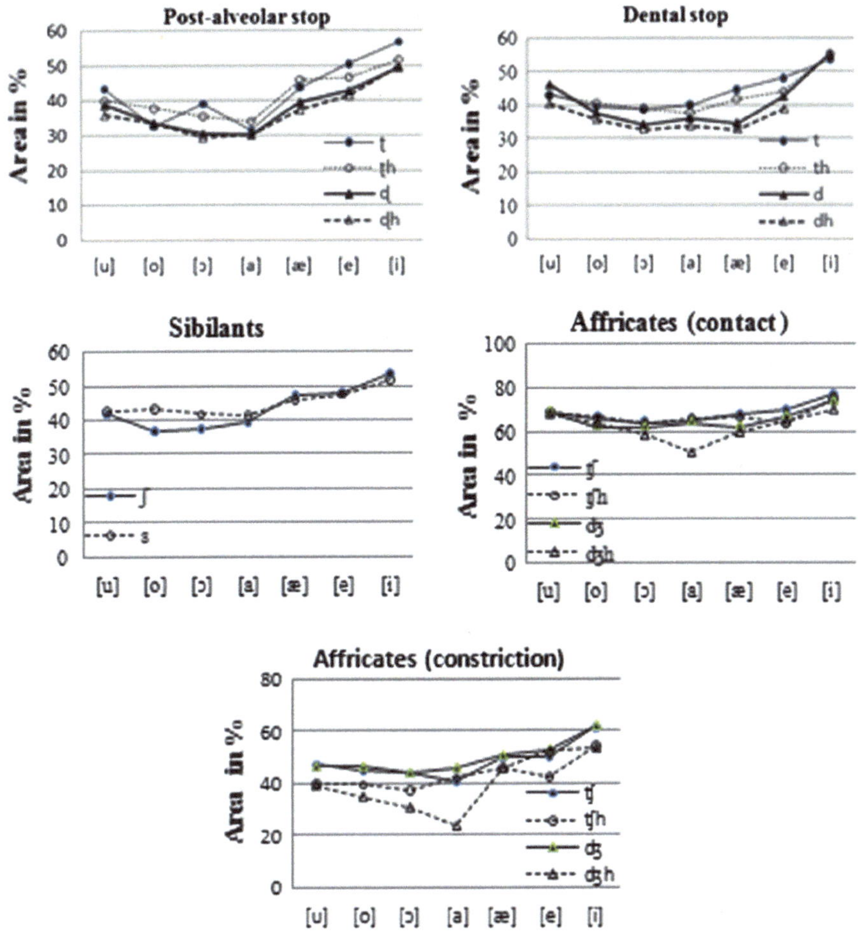

Fig. 2.102 The area of contact/constriction in the oral tract for different consonants

tongue for front leaving lesser time for the consonantal contact. As far as the articulatory positions of stops are concerned both the magnitude and the range of contact area increases from back to front. Area of constriction for affricates is consistently lower than that for firm closure.

2.5.8 *Summary*

For plosive/stops there is a general indication that contact region gets generally forwarded on release by about one row, which is not very insignificant. It indicates that the closure is released by a forward movement of the tongue. For retroflex place of release advances two rows taking it to Post alveolar region from a mideo-palatal closure. The dependence of the place of articulation with respect to sex is consistent but quite moderate, generally forwarded for the female speaker. The influence of vowels on place of release and closure is significant. It is monotonically forwarded from back to front, generally two rows, most backward for /u/ and most forward for /i/.

It may be noticed from Fig. 2.1 that in EPG frame configuration a single row difference may sometime indicate different category in articulation. Therefore identifying plosives/stops by the present system of fine categorization may sometimes be misleading. For other sounds the problem is not significant.

The EPG provides for the first time a vivid visual presentation of the dynamic process of the tongue-body interaction in the articulation of consonants. Though it is intermittent it provides an unambiguous picture within its stated limitation. This provides a new understanding that the place of articulation is not so crisp a categorization as one would like to believe from the traditional description and IPA symbolic representation. In fact the fuzziness is so large in the case of retroflex plosives/stops that one has to reconsider how to define them. The release is at one place and the closure is at a different categorical place. Also the extent of influence of vowels on deciding the place category is too significant to be wished away. A wholesome picture would come out only when this type of study is done with larger number of informants.

One of the important shortcomings of the EPG system is that it does not provide firm contact evidence for velar and dental region. The other one is that as of now it is difficult to have studies done with a large number of speakers. The false palate has to made outside India increasing the cost and time. Once this is made in India larger number of informants could be used. I am sure necessary technological is available in the country. Only one has to tap them.

References

Andruski JE, Ratliff M (2000) Phonation types in production of phonological tone the case of Green Mong. J Int Phon Assoc 30:39–62

Berkins S, Stevens KN (1982) Across language study of the perception of nasal vowels. J Acoust Am 73(suppl 1):s54

Bhattacharya T (2000) Bangla (Bengali). In: Gary J, Rubino C (eds) Encyclopaedia of world's languages: past and present. WW Wilson, New York

Bickley C (1982) Acoustic analysis and perception of breathy vowels. In: Speech communication group working papers, pp 71–82. Cambridge Massachusetts Institute of Technology

Catford JC (1977) Fundamental problems in phonetics. Indiana University Press

Chatterji SK (1926) The origin and development of the Bengali language. Rupa & Co., New Delhi

Chávez-Peón ME (2013) Non-modal phonation in Quiaviní Zapotec: an acoustic investigation. Instituto de Investigaciones Antropológicas, Universidad Nacional Autónoma de México. Retrieved 26 May 2013

Choudhury S (2006) Concatenative text to speech synthesis: a study of Standard Colloquial Bengali. PhD Thesis, Indian Statistical Institute

Colin PM (1991) Cited in Indo-Aryan languages. Cambridge University Press
Crystal D (2003) A dictionary of linguistics & phonetics (5th ed). Wiley-Blackwell
Das Mandal SK (2007) Role of shape domain parameters in speech recognition: a study on Standard Colloquial Bangla (SCB), Ph.D Thesis, Jadavpur University
Datta AK (1988) Acoustic phonetics of non-nasal standard Bengali vowels: a spectrographic study. JIETE 34
Datta AK (2014) Aspirated vowels in Standard Colloquial Bengali: a case study with native informants' communicated to JPH, Springer, Berlin
Datta AK, Ganguly NR (1981) Terminal frequencies in CV Combination in multisyllabic words. Acustica 47(4):314–324
Datta AK, Ganguly NR (1985) Behaviour of terminal frequencies in VC combination. Acustica 67:26–33
Datta AK, Mukherjee B (2011) On the role of formants in the cognition of vowels and place of articulation of plosives. In: Ystad S, Aramaki M, Kronland-Marinet R, Mohanty S (eds) Speech, sound and music processing: embracing research in India
Datta AK, Dutta Majumder D, Ganguly NR, Mukherjee B, Sarkar R (1974) Studies on acoustic phonetic features of Telugu speech sounds. ECSL series on phonological studies, Nov 1974
Datta AK, Ganguli NR, Ray S (1978a) Transition—a cue for identification of plosives. J Acoust India VI(4):124–131
Datta AK, Ganguly NR, Dutta Majumder D (1978b) Some studies on acoustic features of Telugu vowels. Acustica 41:55–64
Datta AK, Ganguly NR, Ray S (1980) Recognition of unaspirated plosives: a statistical approach. IEEE Trans Acous Speech Sig Process ASSP-28(1):85–91
Datta AK, Ganguly NR, Dutta Majumder D (1981) Acoustic features of consonants; a study based on Telugu speech sounds. Acustica 47(2)
Datta AK, Ganguly NR, Mukherjee B (1989) Bengali nasal sounds—a spectrographic study. J Acous Soc India XVII:219–223
Datta AK, Sengupta R, Dey N, Banerjee BM, Nag D (1998) Perception of nasality in Bengali vowels: role of harmonics between F0 and F1. In: Proceedings of international conference of computational linguistics, speech and document processing, ISI, Calcutta, Feb 18–20, 1998
Datta AK, Sengupta R, Dey N (2003) Jitter, Shimmer and HNR characteristics of singers and non-singers. J ITC Sangeet Res Acad 17
Delattre PC, Liberman AM, Cooper FS, Gerstman LJ (1952) An experimental study of the acoustic determinants of vowel colour; observations on one- and two-formant vowels synthesized from spectrographic patterns. Word 8(3):195–210
Dioubina OI, Pfitzinger HR (2002) An IPA vowel diagram approach to analysing L1 effects on vowel production and perception. In Proceedings of ICSLP '02, Denver, vol 4, pp 2265–2268
Djordje K, Das R (1972) Short outline of Bengali Phonetics. Statistical Publishing Society, Calcutta
Dutta Majumder D, Datta AK (1966) A scheme for automatic speech coding and recognition. In: ISALS symposium on control computation
Esposito CM, Khan SUD, Hurst A (2005) A breathy nasals and /Nh/Clusters in Bengali, Hindi, and Marathi. UCLA working papers in phonetics, vol 104, pp 82–106
Fant G (1970) Acoustic theory of speech production. Mouton De Gruyter
Fant G, Ishizaka K, Lindquist-Gauffin J, Sundberg J (1972) Subglottal formants, STL-QPSR 1/1972
Fischer-Jorgensen E (1967) Phonetic analysis of breathy (murmured) vowels in Gujarati. Indian Linguist 28:71–139
Fujimara O (1960) Spectra of nasalised vowels. Res Lab Electrn Q Prog Rep No. 58, MIT 214–218
Ganguli NR, Datta AK, Mukherjee B (1988) Acoustic phonetics of non-nasal Standard Bengali Vowels: a spectrographic study. J Electron Telecommun Eng 34(1):50–56
Ganguly NR, Datta AK, Mukherjee B (1999) Acoustic phonetic features of glides and dipthongs in Bengali. J Acous Soc India XXVII:199–202

Gordon M (2001) Phonation types: a cross-linguistic overview. J Phon
Hai AM (1989) Dhani Bigyan and Bangla Dhani Tattwa. Mallick Brothers, Dhaka, Bangladesh
Hanson H (1995) Glottal characteristics of female speakers, Ph.D. dissertation. Harvard University, MA
Hartmut RP (2003) Acoustic correlates of the IPA vowel diagram. 15 ICPhS Barcelona, pp 1441–1444
Hawkins S, Stenens KN (1985) Acoustic and perceptual correlates of the non-nasal-nasal distinction for vowels. J Acoust Am 77(4):1560
Heinz JM, Stevens KN (1961) On the properties of voiceless fricative consonants. J Acoust Am 33:589–596
Hideaki K, Hideo K, Jun T, Masaru S (2006) Spectral properties of Japanese whispered vowels referred to pitch. J Acoust Soc Am 120(5):3378–3378
Hillenbrand J, Houde RA (1996) Acoustic correlates of breathy vocal: quality dysphonic voices and continuous speech. J Speech Hear Res 39:311–321
Hillenbrand J, Cleveland RA, Erickson RL (1994) Acoustic correlates of breathy vocal quality. J Speech Hear Res 37:769–778
Hombert JM, Ohala J, Ewan W (1979) Phonetic explanations for the development of tones. Language 55(1):37–58
Hossain SA, Rahman ML, Ahmed F (2005) Acoustic space of Bangla vowels. In: WSEAS 5th international conference on speech and image processing, Greece, pp 138–140, August 2005
Huffman MK (1987) Measures of phonation types in Hmong. J Acoust Soc Am 81(1):495–504
Ioana C (2002) A perception-production study of Romanian diphthongs and glide-vowel sequences. J Int Phon Assoc 32(2):203–221
Jones D (1962) An outline of English phonetics. W. Heffer & Sons Ltd., Cambridge
Klatt DH, Klatt C (1990) Analysis, synthesis, and perception of voice quality variations among female and male talkers. J Acoust Soc Am 87(2):820–857
Ladefoged P (1967) Three areas of experimental phonetics. Oxford University Press, Oxford
Ladefoged P, Antananzas-Barroso N (1985) Computer measures of breathy voice quality. UCLA working papers in phonetics, vol 61, pp 79–86
Ladefoged P, Maddieson I (1996) The sounds of the world's languages. Blackwell, Oxford Cambridge, MA
Laver J (1980) The phonetic description of voice quality. Cambridge University Press
Lehiste I, Peterson EG (1961) Transitions, glides, and diphthongs. JASA 33(3):268–277
McGowan RS (1922) Tongue-tip trills and vocal-tract wall compliance. JASA 91:2903–2910
Mistry PJ (1997) Gujarati phonology. In: Kaye AS (ed) Phonologies of Asia and Africa. Winona Lake Eisenbrauns
Ngoc YPT, Badin P (1994) Vocal tract acoustic transfer function measurements further developments and applications. Suppliment au Journal de Physique 111, vol 4, May 1994
Padgett J (2007) Glides, vowels, and features. Lingua 118(12):1937–1955
Pandit PB (1954) Indo Aryan sibilants in Gujarati. IL, 14
Perkell JS (1969) Physiology of speech production: results and implications of a quantitative cineradiographic study. Research Monograph No. 53, MIT Press, Cambridge MA
Recasens P, Pallare MD (1999) A study of /Q/ and /r/ in the light of the "DAC" coarticulation model. J Phon 27:143–169
Saksena BR (1971) *Evolution of Awadi*. Motilal Banarasi Das Publication, New Delhi, pp 74–76
Sarkar P (1985) Bangla Dwishardhani (Bangla Diphthong), Calcutta, 1985–86
Schane S (1995) Diphthongization in particle phonology. In: Goldsmith JA (ed) The handbook of phonological theory. Blackwell handbooks in linguistics. Blackwell, pp 586–608
Shaddle CH (1990) Articulatory-acoustic relationships in fricative consonants. In: Hardcastle WJ, Marchal A (eds) Speech production and speech modelling, Kulwer, Dordrecht, The Netherlands, pp 187–209
Spajić S, Ladefoged P, Bhaskararao P (1996) The trills of Toda. J Int Phon Assoc 26(1):1–22
Stevens K (2000) Acoustic phonetics. MIT Press, Massachusetts

Stevens K, Hanson H (1994) Classification of glottal vibration from acoustic measurements. Presented at the 8th vocal fold physiology conference, Kurume, Japan, April 7–9, 1994

Takeuchi S, Kasuya H, Kido K (1975) On the acoustic correlates of nasality. J Acoust Jpn 31: 298–309

Tartter VC (1989) What's in a whisper? J Acoust Soc Am 86(5):1678–1683

Teagre HM, Teagre SM (1990) A phenomenological model for vowel production in the vocal tract. In: Daniloff RG (ed) Speech science recent advances, College Hill, San Diago, pp 73–109

Thongkum T (1988) Phonation types in Mon-Khmer languages. In: Fujimura O(ed) Vocal fold physiology voice production, mechanisms and functions. Raven Press, New York, pp. 319–334

Tsunoda K, Ohta Y, Soda Y, Niimi S, Hirose H (1997) Laryngeal adjustment in whispering: magnetic resonance imaging study. Ann Otol Rhinol Laryngol 106:41–43

Varshney RL (1995) An introductory textbook of linguistics and phonetics, 8th edn. Student Publication store

Wayland R, Jongman A (2003) Acoustic correlates of breathy and clear vowels the case of Khmer. J Phon 31:181–201

http://www.cdackolkata.in/html/txttospeeh/corpora/corpora_main/MainB.html

http://www.phonetics.ucla.edu/course/chapter1/vowels.html

http://www.speech.kth.se/wavesurfer

Epilogue

The entire work presented in this book has the underlying motivation to have objective data and objective categorization on Bangla consonants which represents the contemporary speech of native Bangla speakers in relation to the subjective traditional description of scholars. The book attempts to consolidate the substances of investigations carried out over about last three decades in different organizations, particularly in Indian Statistical Institute, Kolkata and CDAC, Kolkata. It may be quite possible that some work in some place may have been overlooked. If so, it is regretted.

While manner based categorization seem to be relatively consistent and reasonably robust, that for the place of articulation is somewhat fuzzy. As speech is a dynamically evolving subject, the traditional observations and categorizations need to be evaluated from time to time. The number of speakers and the way the data has been collected seem to be just adequate for representation of what is actually spoken now as far as the acoustics is concerned. Of course it is true that more number of informants, number of words, and use of really free speech would provide better representation. This will, hopefully, be taken up in the future. On this respect the present EPG analysis is really handicapped. The EPG analysis has been conducted only on one informant of each sex because of the logistic constraints. One of them is the cost of making the palate, which has to be done from abroad. The differences observed between sexes are therefore only indicative. One wishes that the entire analysis may be done some time over a comparatively larger number of informants.

But a more fundamental issue needs to be at least broached here. It is generally said that for the development of technology precise objective parametric representations are needed. When one talks about precision one does not necessarily mean deterministic precision. It could very well be probabilistic with appropriately defined statistics. Speech technologies, particularly the ASR and TTS technologies have to deal with high dose of subjectivity. Even the paradigm of stochastic determinism seems to be neither adequate nor appropriate. In speech the subject is too much entangled with the process of producing some objective reality, say speech or songs. For example if a Bangla native speaker says /pin/ it is likely to be heard as /bin/ by a native listener of London. Similarly the word 'king' spoken by an Englishman is likely to be heard as /khiŋ/ by a Bangla listener. There are many

A.K. Datta, *Acoustics of Bangla Speech Sounds*, Signals and Communication Technology, DOI 10.1007/978-981-10-4262-1 such instances. In fact it is said that there is about 40% of objective content in the form of acoustics in the speech signal. The rest 60% of the knowledge to decipher the oral message come from the listener's brain. The acoustic parameters are affected not only by the raw abstract phones but by many other highly semiotic factors like syntax, semantics, pragmatics etc. Until one has the proper paradigm one has to redefine the objectivity from simple pointer reading to a process of quantification where the subjectivity is embraced deeply. One can only hope the variations induced by subjects could be modeled by statistics.

Speech is a natural intelligent process. The context involved in the interpretation of speech is also an intelligent context not merely a syntactic context. One would need subjectively objective measures which can quantify sensory inputs like sound waves into subjective evaluation of the objective fundamentals. This is nothing new. Some such measures already exist e.g., sone, pitch, etc. It may be pertinent here to mention the recent experiment of Hartmut (2003) establishing a relationship between objective spectral characteristics with the subjective cognitive classification in the case of vowels. These involve psycho perceptual studies involving large number of informants. One such experiment was reported here by the present authors in the case of nasality of vowels. Such experiments need to be conducted for all sound classes in the speech domain. This is likely to plug the confusion between cognition and acoustics.

Be that as it may, one cannot sit idle till an appropriate paradigm appears on the horizon. The development of technology cannot wait. One needs to use properly the best technologies, including instruments and methodologies, which are available for producing as a good a data set as possible which represent the speech and hearing of native subjects. The researchers in Indian Statistical Institute and CDAC, Kolkata tried painstakingly to evolve such a format and necessary paradigm over the years. The present work emulates those. One may note here that in the last decade acoustic phonetic data in a number of standard regional dialects are being collected. Unfortunately they are dispersed at many places. Moreover each has their own formats which are quite varied. It is necessary that a standard format and paradigm is evolved for collection and presentation of such data. The author is not aware of whether such a paradigm for Indian spoken languages exists.

The last but not the least is a few words about the impact of a book like this in the knowledge domain of the nation. As far as the author believes that the students and young researchers of linguistics in India is in serious need of a source book which provides comprehensive details of speech acoustic, particularly of a Indian dialect. The author hopes that this book may be useful for them.

Reference

Hartmut RP (2003) Acoustic correlates of the IPA vowel diagram. In: 15 ICPhS Barcelona, pp 1441–1444

Appendix

See Figs. A.1 and A.2.

Place of release generally forwarded for female.

Place of release forwarded for back vowels irrespective of sex.

See Figs. A.3, A.4, A.5, A.6, A.7, A.8, A.9, A.10, A.11 and A.12.

For plosives [k, k^h, g, g^h,] place of articulation generally forwarded for female and for back vowels irrespective of sex.

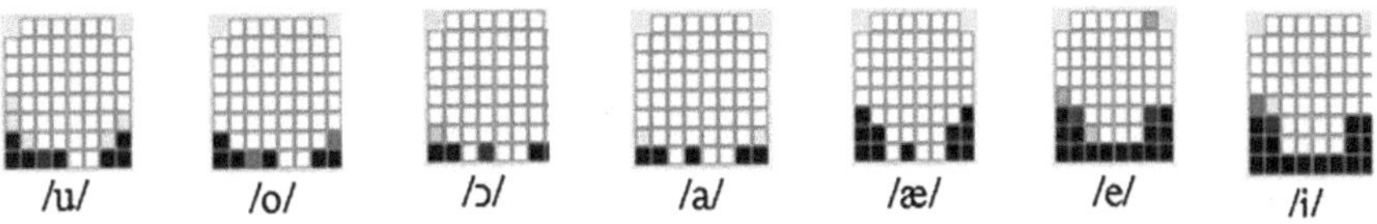

Fig. A.1 Place of release of [k^h] for the Bangla vowels for the female speaker

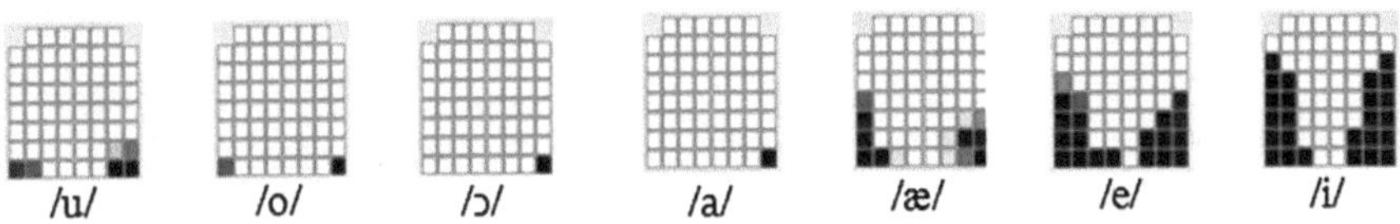

Fig. A.2 Place of release of [k^h] for the Bangla vowels for the male speaker

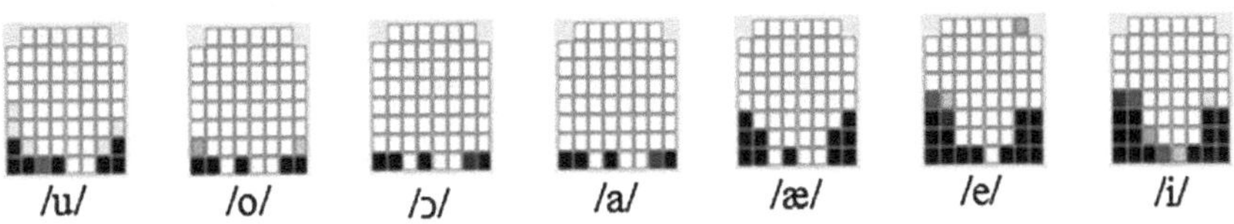

Fig. A.3 Place of closure of [k^h] for the Bangla vowels for the female speaker

A.K. Datta, *Acoustics of Bangla Speech Sounds*, Signals and Communication Technology, DOI 10.1007/978-981-10-4262-1

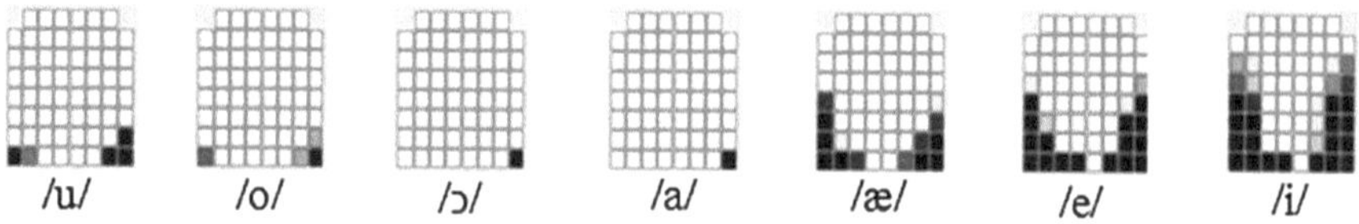

Fig. A.4 Place of closure of [k^h] for the Bangla vowels for the male speaker

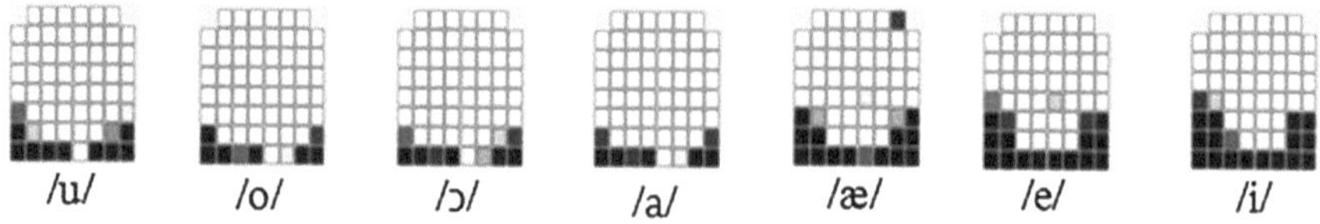

Fig. A.5 Place of release of [g] for the Bangla vowels for the female speaker

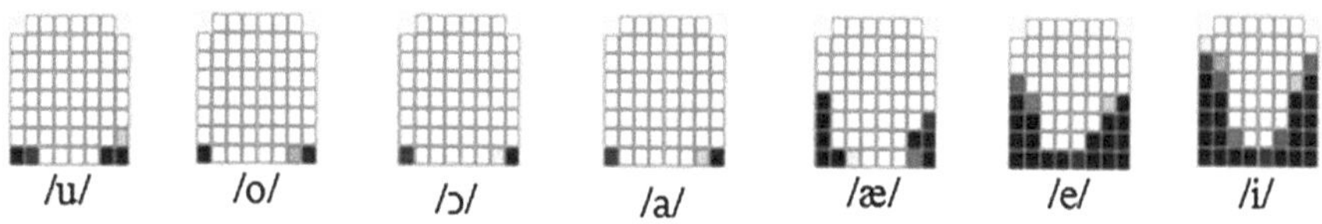

Fig. A.6 Place of release of [g] for the Bangla vowels for the male speaker

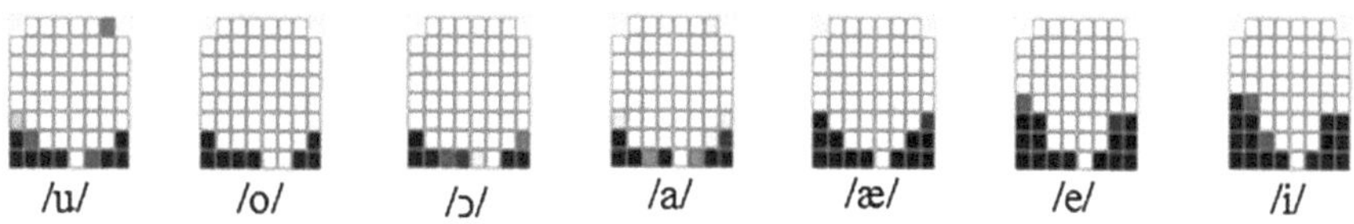

Fig. A.7 Place of closure of [g] for the Bangla vowels for the female speaker

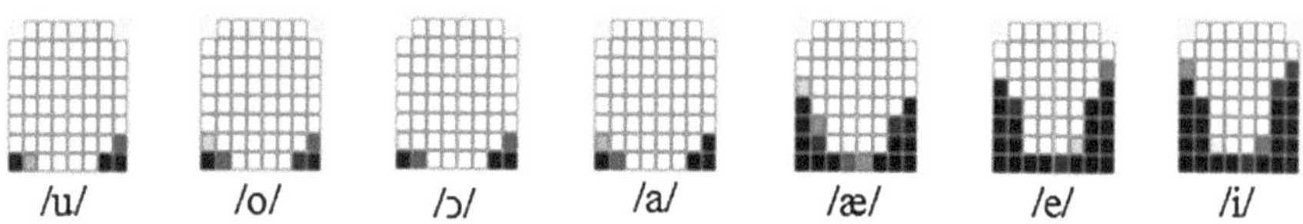

Fig. A.8 Place of closure of [g] for the Bangla vowels for the male speaker

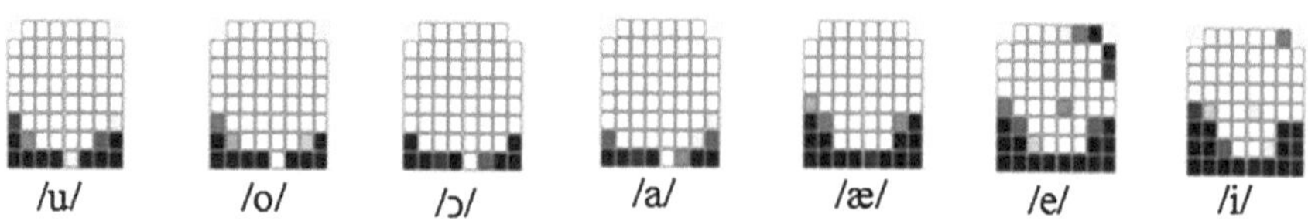

Fig. A.9 Place of release of [g^h] for the Bangla vowels for the female speaker

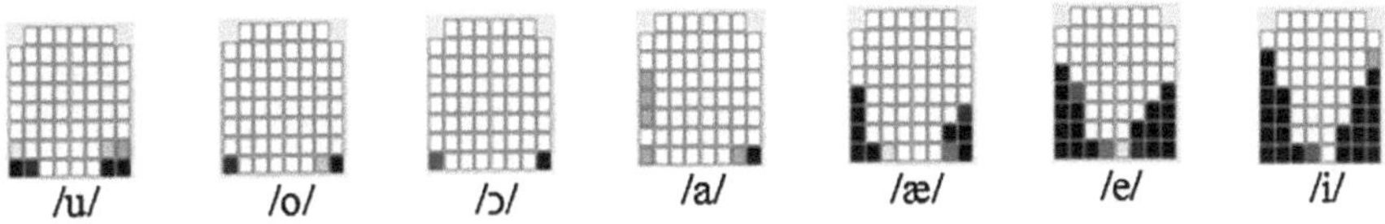

Fig. A.10 Place of release of $[g^h]$ for the Bangla vowels for the male speaker

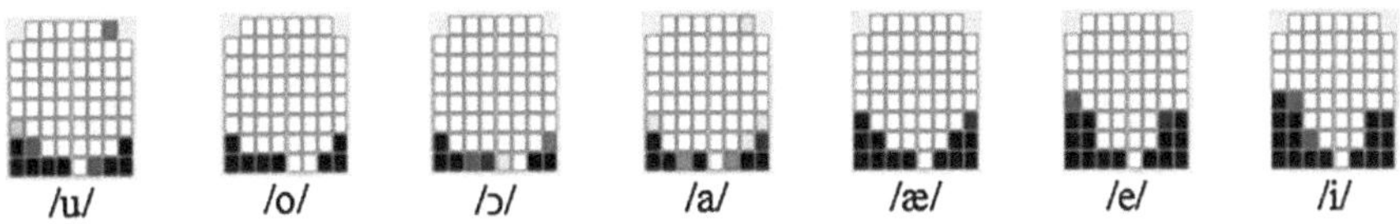

Fig. A.11 Place of closure of $[g^h]$ for the Bangla vowels for the female speaker

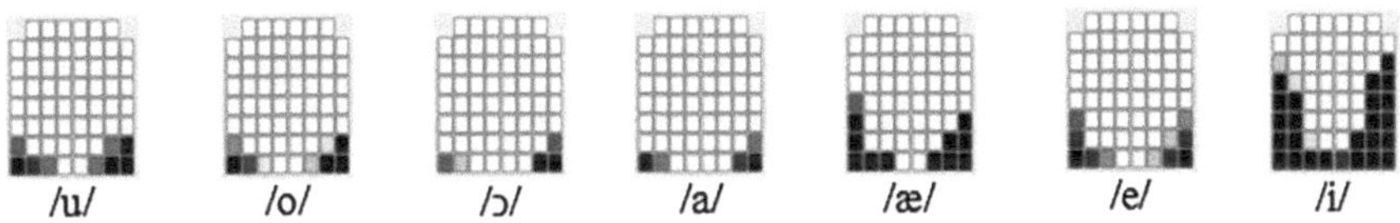

Fig. A.12 Place of closure of $[g^h]$ for the Bangla vowels for the male speaker

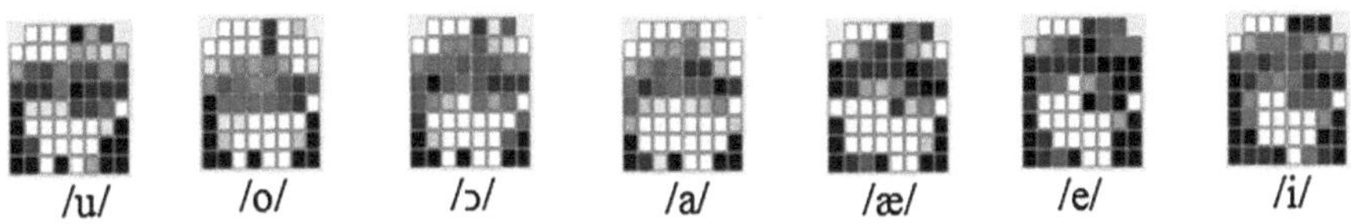

Fig. A.13 Place of release for [ʈ] for all the Bangla vowels for the female speaker

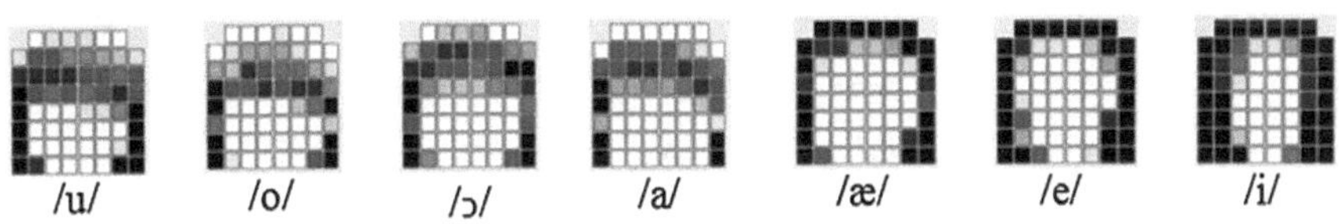

Fig. A.14 Place of release for [ʈ] for all the Bangla vowels for the male speaker

Figures A.13, A.14, A.15, A.16, A.17, A.18, A.19, A.20, A.21, A.22, A.23, A.24, A.25, A.26, A.27 and A.28 reveal that the place of release for all retroflex plosives is post alveolar for the back vowels for both male and female speakers. For front vowels the tongue tip is forwarded and the release takes place almost at the root of teeth for male speaker. Also for female speakers contact place gets forwarded but only marginally. The place of closure is post alveolar in case of front

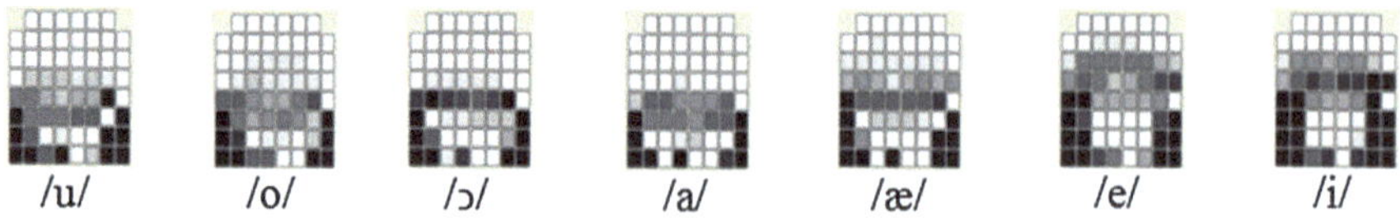

Fig. A.15 Place of closure for [ʈ] for all the Bangla vowels for the female speaker

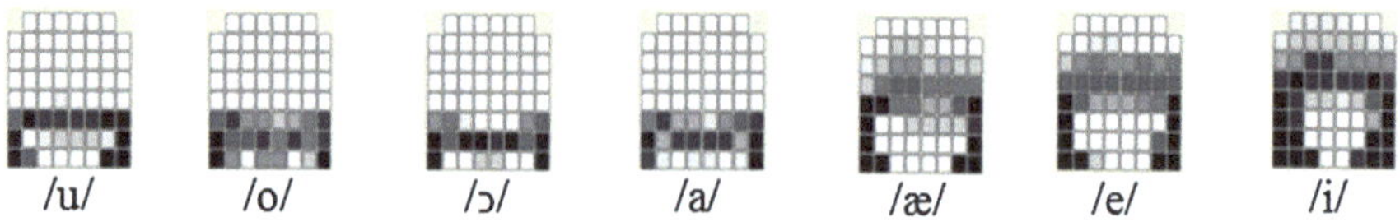

Fig. A.16 Place of closure for [ʈ] for all the Bangla vowels for the male speaker

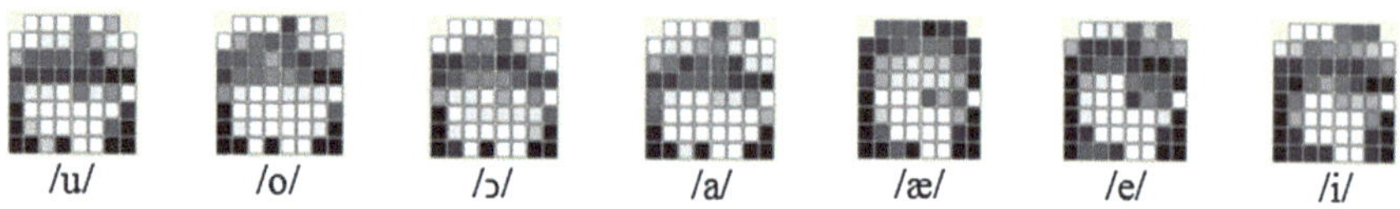

Fig. A.17 Place of release of [ʈ^{h}] for the all Bangla vowel for the female speaker

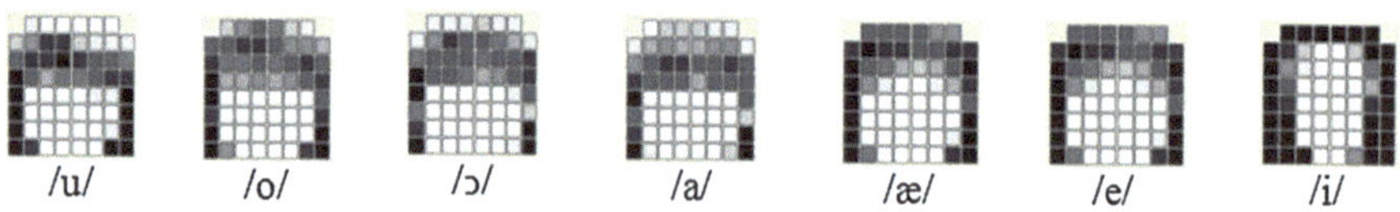

Fig. A.18 Place of release of [ʈ^{h}] for all the Bangla vowels for the male speaker

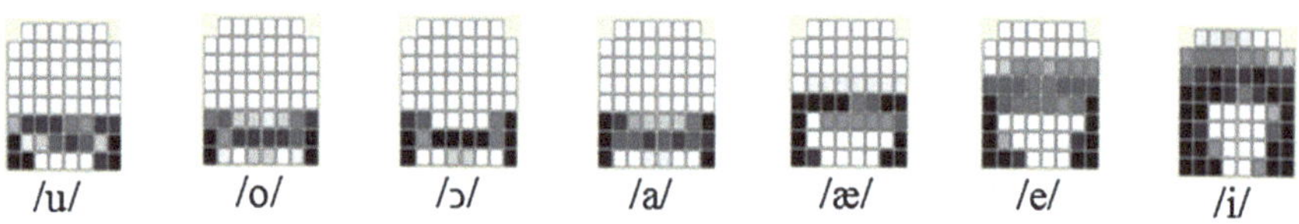

Fig. A.19 Place of closure of [ʈ^{h}] for all the Bangla vowels for the female speaker

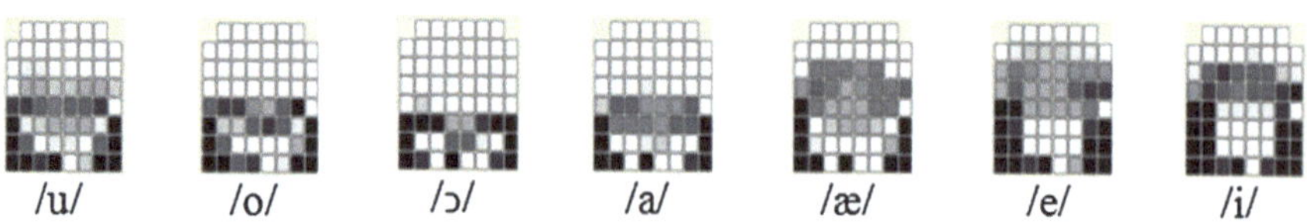

Fig. A.20 Place of closure of [ʈ^{h}] for all the Bangla vowels for the male speaker

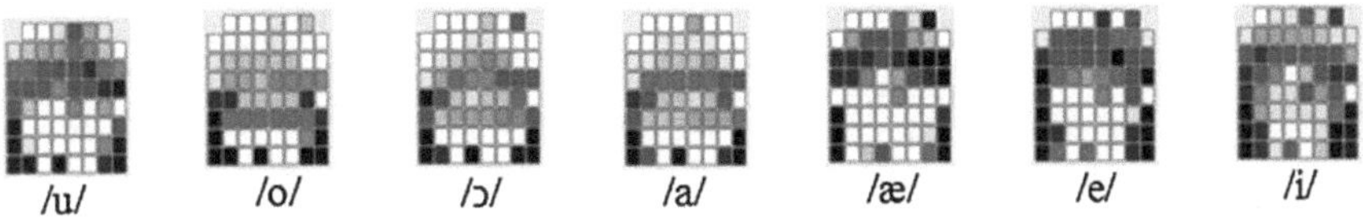

Fig. A.21 Place of release of [ɖ] for all the Bangla vowels for the female speaker

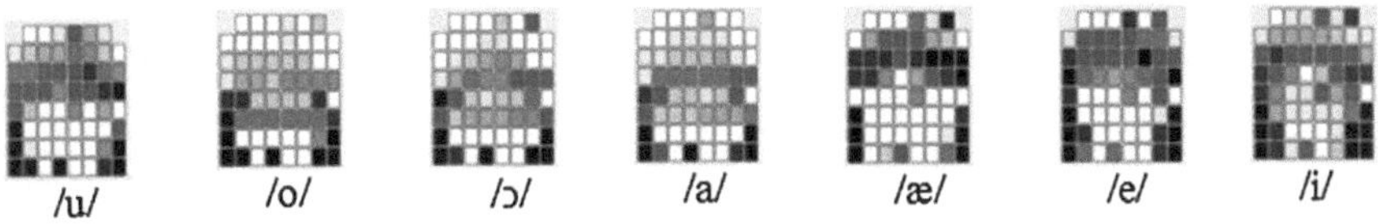

Fig. A.22 Place of release of [ɖ] for all the Bangla vowels for the male speaker

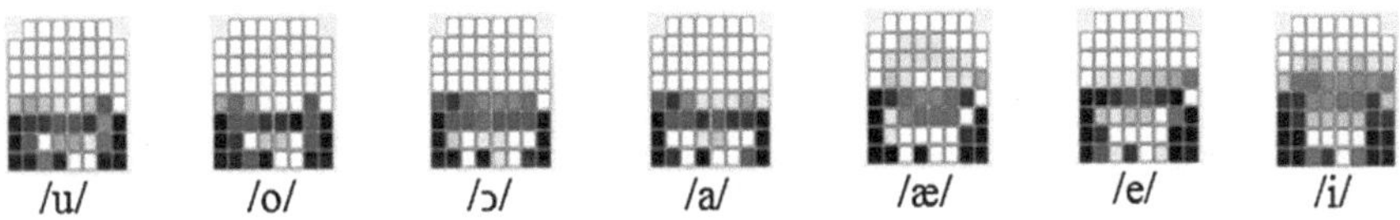

Fig. A.23 Place of closure of [ɖ] for all the Bangla vowels for the female speaker

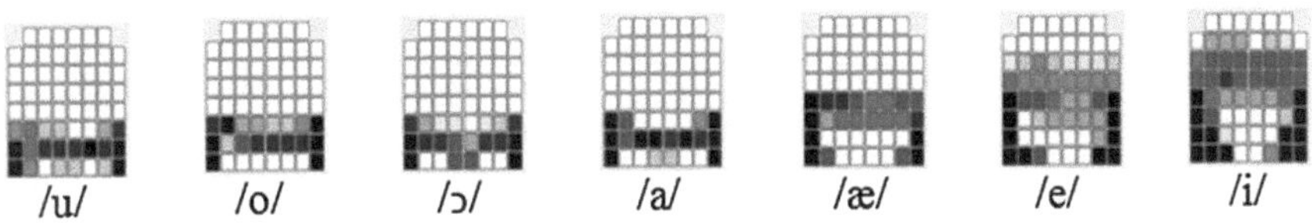

Fig. A.24 Place of closure of [ɖ] for all the Bangla vowels for the male speaker

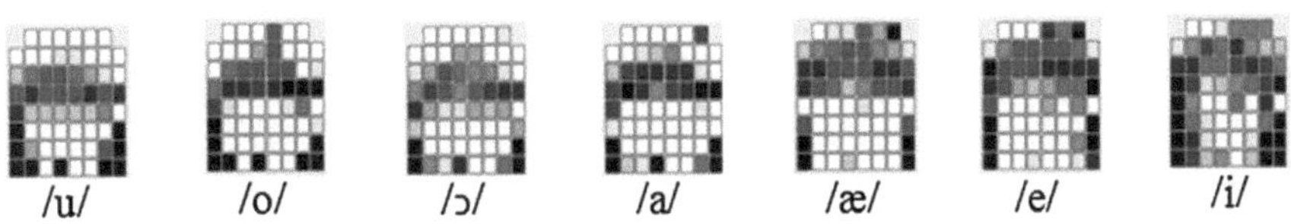

Fig. A.25 Place of release of [$ɖ^h$] for all the Bangla vowels for the female speaker

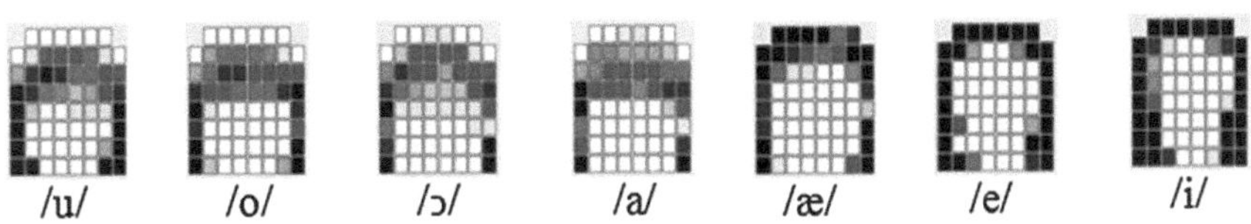

Fig. A.26 Place of release [$ɖ^h$] for all the Bangla vowels for the male speaker

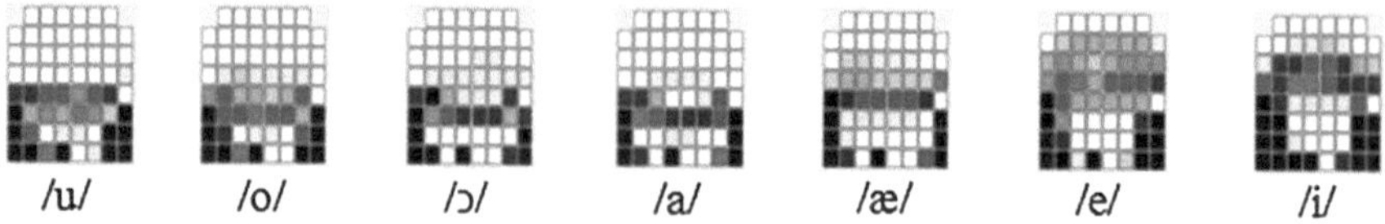

Fig. A.27 Place of closure for [d̪^{h}] with seven Bangla vowels for the female speaker

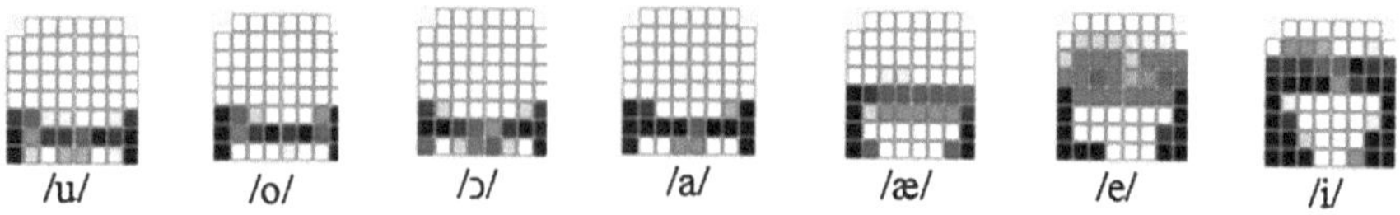

Fig. A.28 Place of closure for [d̪^{h}] with seven Bangla vowels for the male speaker

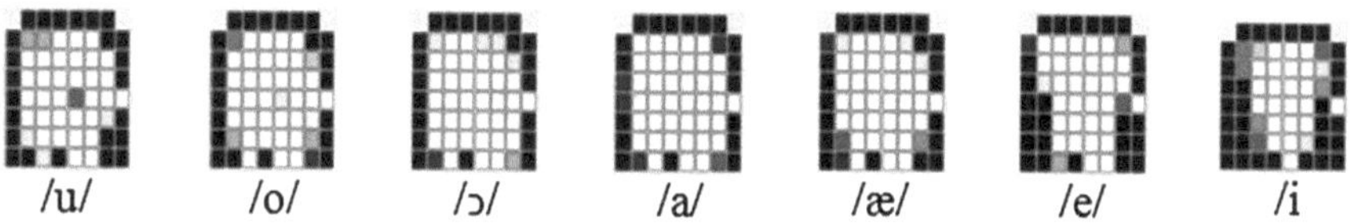

Fig. A.29 Place of release of [t] for all the Bangla vowels for the female speaker

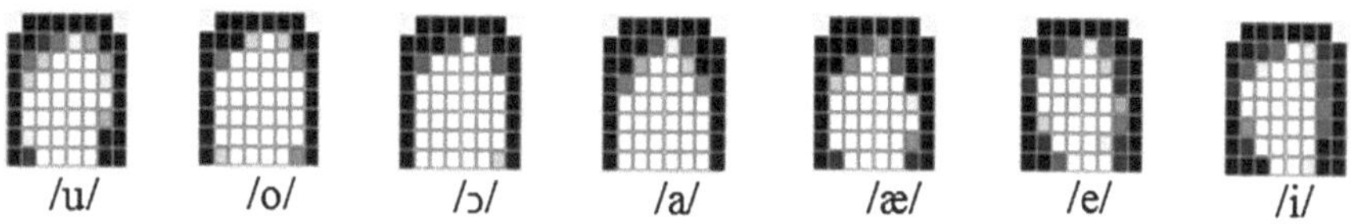

Fig. A.30 Place of release of [t] for all the Bangla vowels for the male speaker

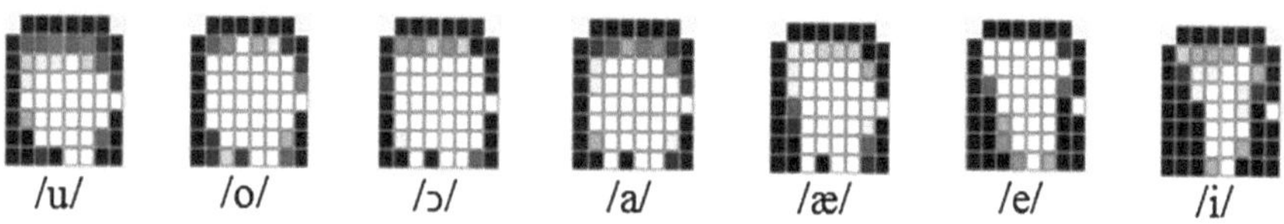

Fig. A.31 Place of closure of [t] for all the Bangla vowels for the female speaker

vowel but it is palatal in case of back vowel for both the speakers. In fact for back vowels the place of closure is mideo-palatal. The closure, in general, may be considered as palatal.

See Figs. A.29, A.30, A.31, A.32, A.33, A.34, A.35, A.36, A.37, A.38, A.39, A.40, A.41, A.42, A.43 and A.44.

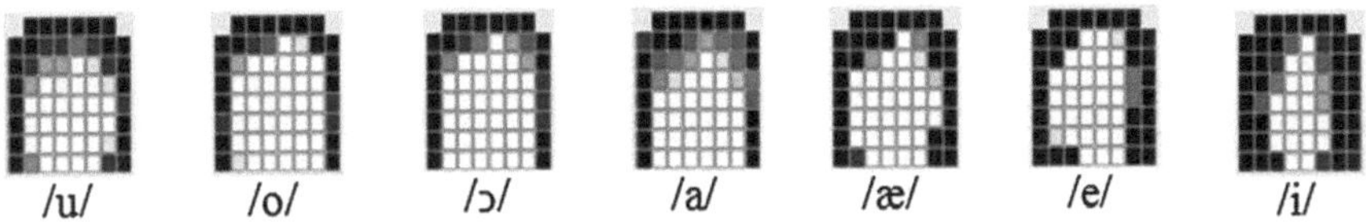

Fig. A.32 Place of closure of [t] for all the Bangla vowels for the male speaker

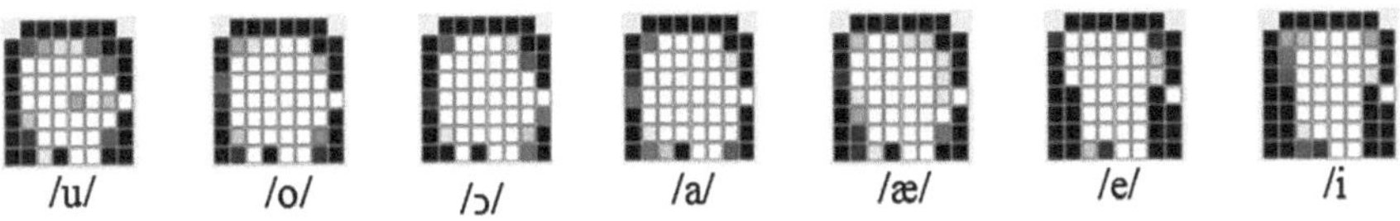

Fig. A.33 Place of release of $[t^h]$ for all the Bangla vowels for the female speaker

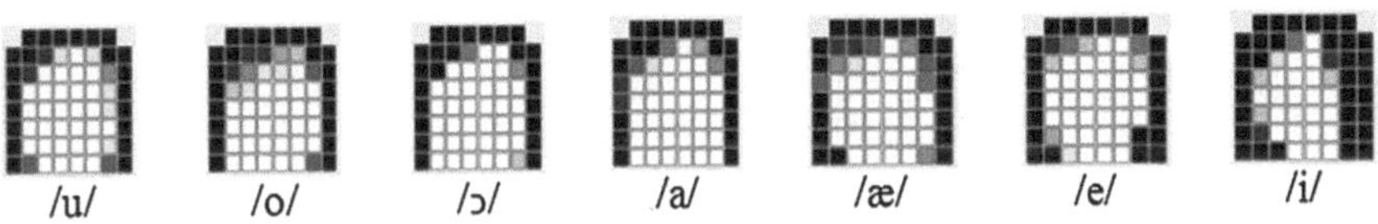

Fig. A.34 Place of release of $[t^h]$ for all the Bangla vowels for the male speaker

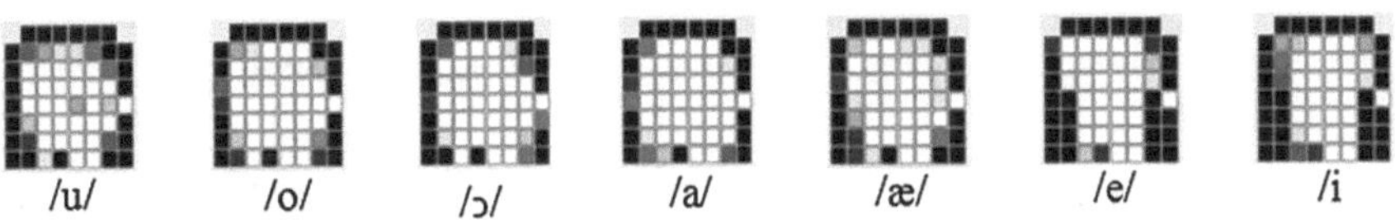

Fig. A.35 Place of closure of $[t^h]$ for all the Bangla vowels for the female speaker

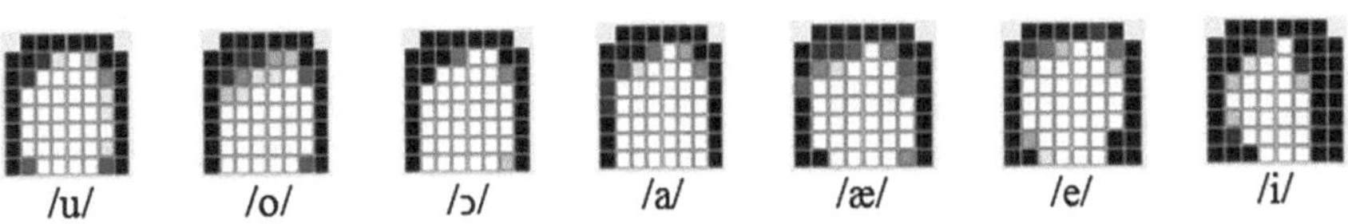

Fig. A.36 Place of closure of $[t^h]$ for all the Bangla vowels for the male speaker

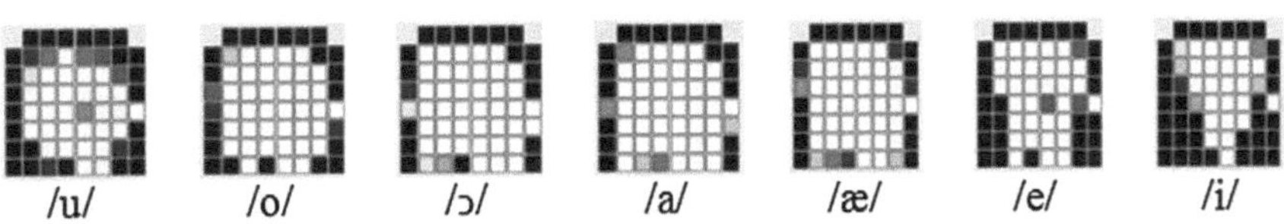

Fig. A.37 Place of release of [d] for all the Bangla vowels for the female speaker

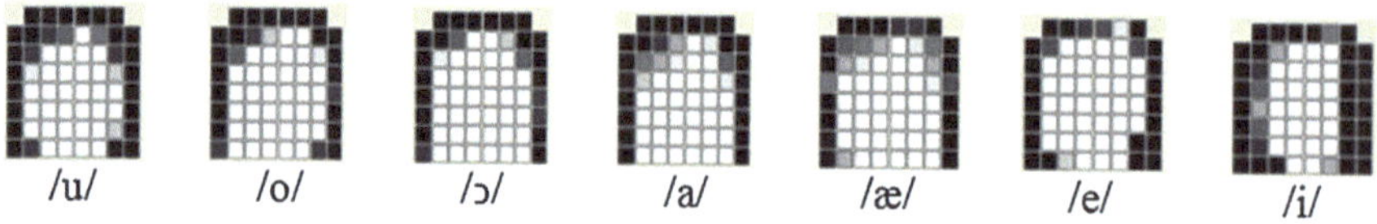

Fig. A.38 Place of release [d] for all the Bangla vowels for the male speaker

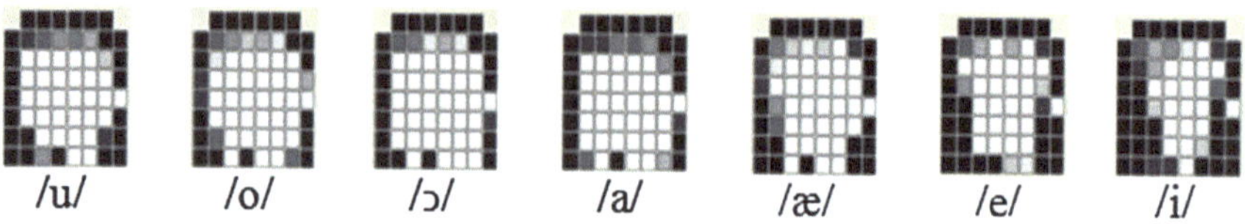

Fig. A.39 Place of closure [d] for all the Bangla vowels for the female speaker

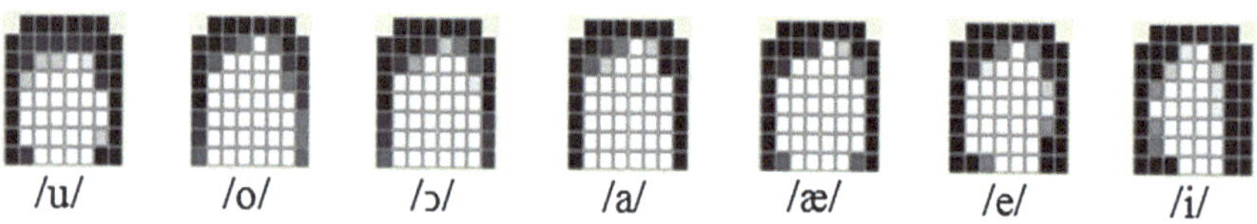

Fig. A.40 Place of closure [d] for all the Bangla vowels for the male speaker

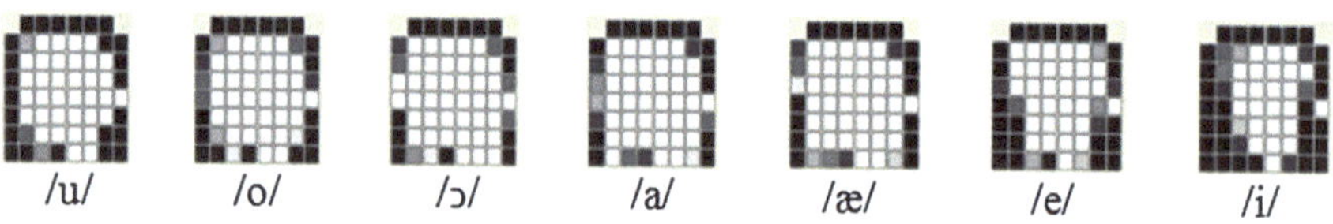

Fig. A.41 Place of release of [d^h] for all the Bangla vowels for the female speaker

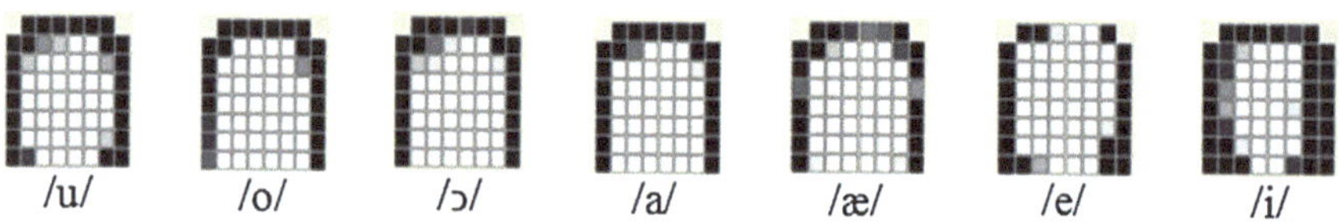

Fig. A.42 Place of release of [d^h] for all the Bangla vowels for the male speaker

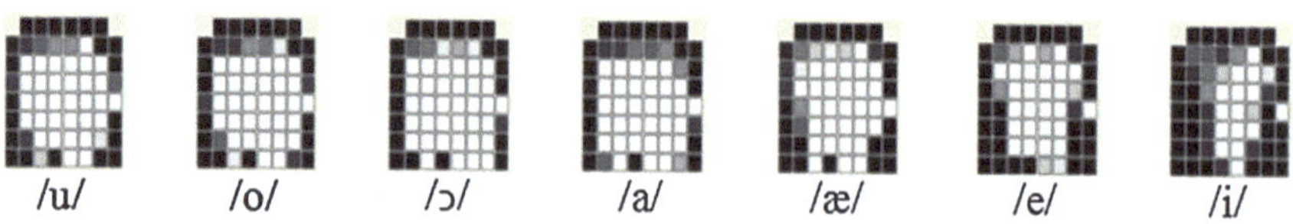

Fig. A.43 Place of closure of [d^h] for all the Bangla vowels for the female speaker

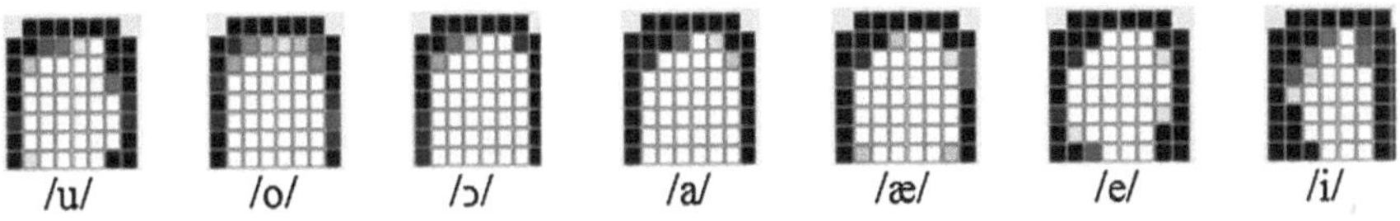

Fig. A.44 Place of closure of [d^h] for all the Bangla vowels for the male speaker

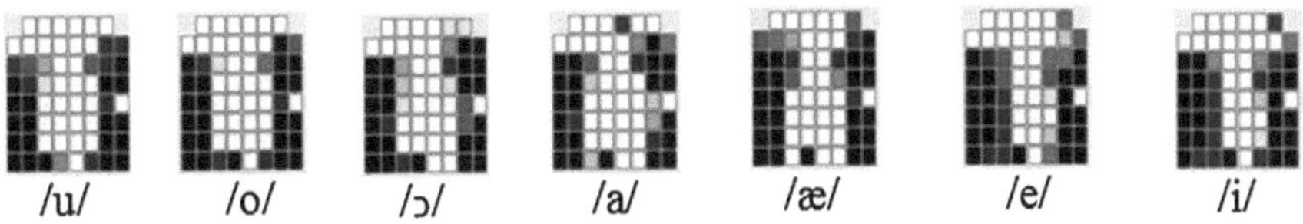

Fig. A.45 Place of constriction of [ʃ] for all the Bangla vowels for the female speaker

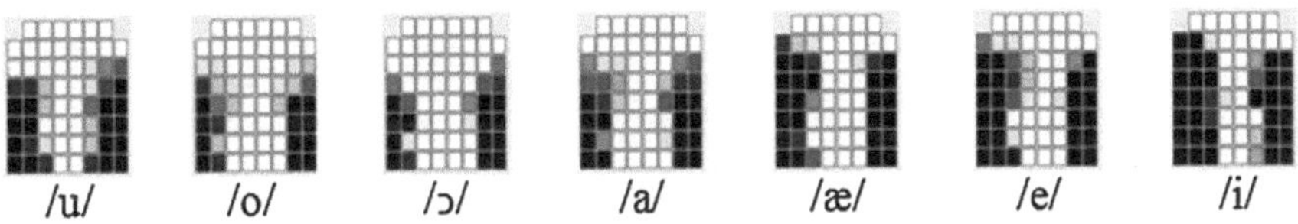

Fig. A.46 Place of constriction of [ʃ] for all the Bangla vowels for the male speaker

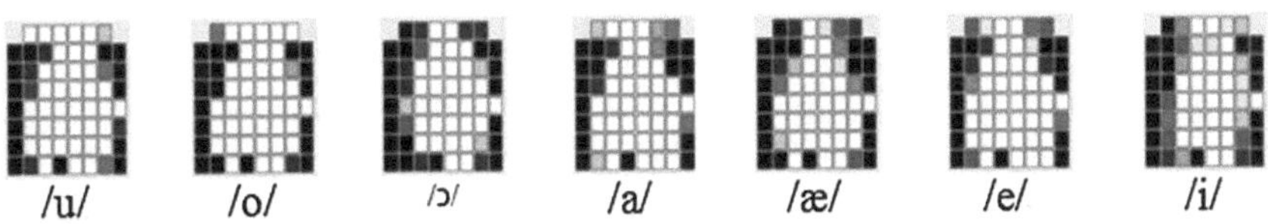

Fig. A.47 Place of constriction of [s] for all the Bangla vowels for the female speaker

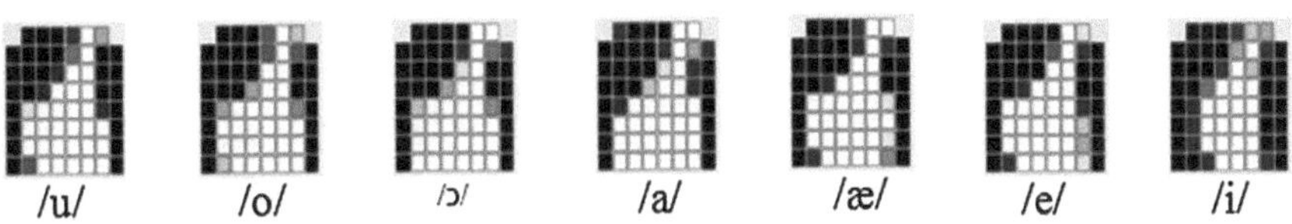

Fig. A.48 Place of constriction of [s] for all the Bangla vowels for the male speaker

For all dental plosives female speaker tongue appears to be a little bit forwarded. Influence of vowels on place of release is not noticeable.

Fricatives
See Figs. A.45, A.46, A.47 and A.48.

Constriction length appears to be a little shorter for back vowels /u,o/

The shape of constriction is an asymmetric funnel for the female speaker almost same for all vowels.

See Figs. A.49, A.50, A.51, A.52, A.53, A.54, A.55 and A.56.

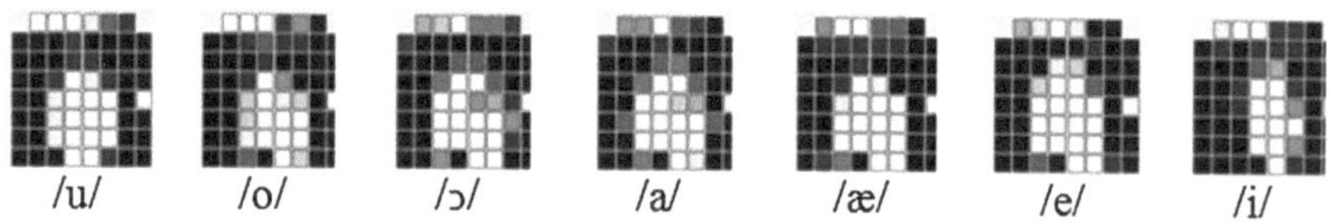

Fig. A.49 place of closure of [ʧ^h] for seven Bangla vowels of female speaker

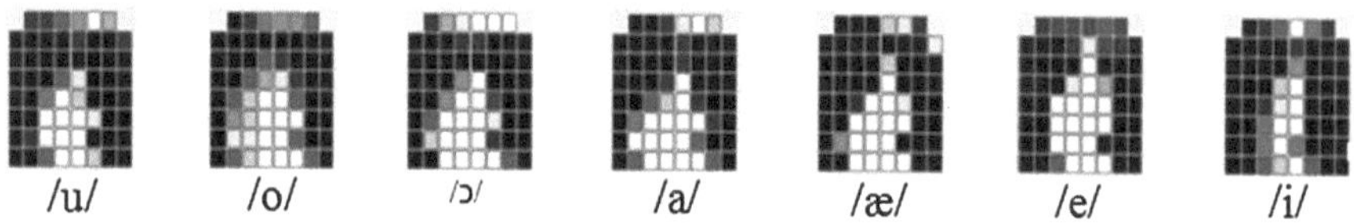

Fig. A.50 Place of closure of [ʧ^h] for seven Bangla vowels for male speaker

Fig. A.51 Place of constriction of [ʧ^h] for seven Bangla vowels for female speaker

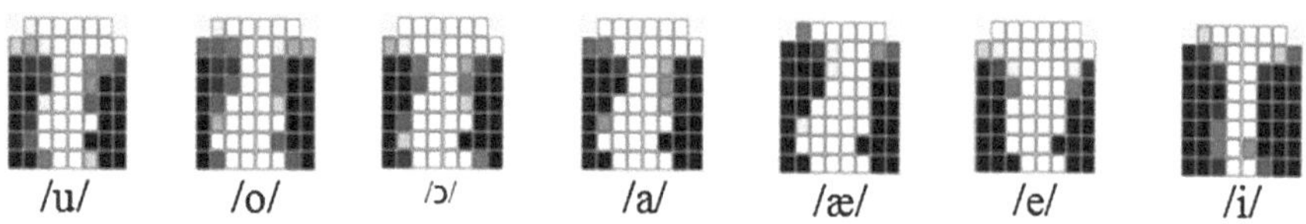

Fig. A.52 Place of constriction of [ʧ^h] for seven Bangla vowels for male speaker

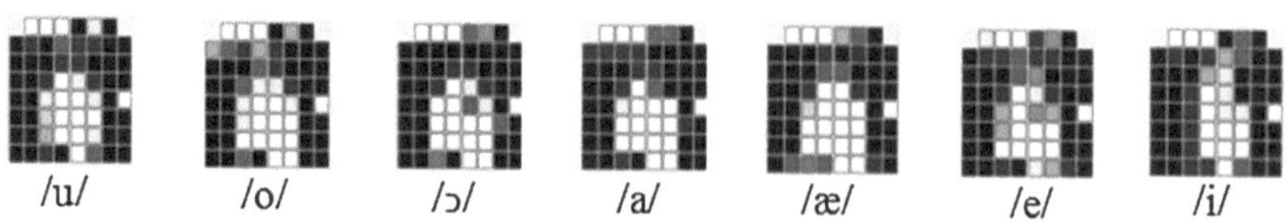

Fig. A.53 Place of closure of [ʤ] for seven Bangla vowels of female speaker

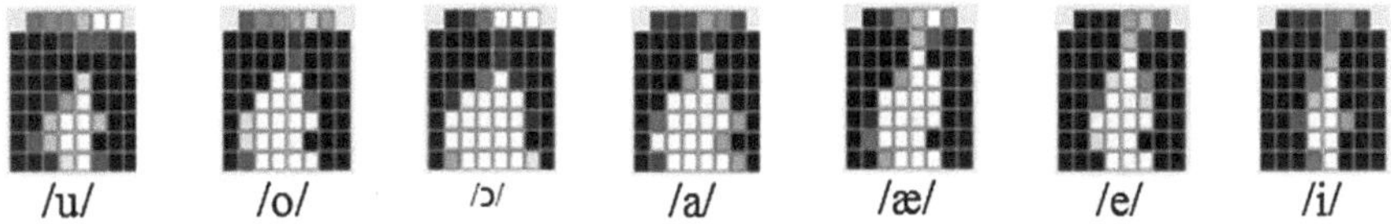

Fig. A.54 Place of closure of [ʤ] for seven Bangla vowels of male speaker

Constriction for male speaker is clearly visible. Comparison of closure frame with those of constriction indicates retraction of tongue on release.

Constriction for male speaker is clearly visible, remarkably long for vowel /i/. Comparison of closure frame with those of constriction indicates retraction of tongue on release.

See Figs. A.57, A.58, A.59 and A.60.

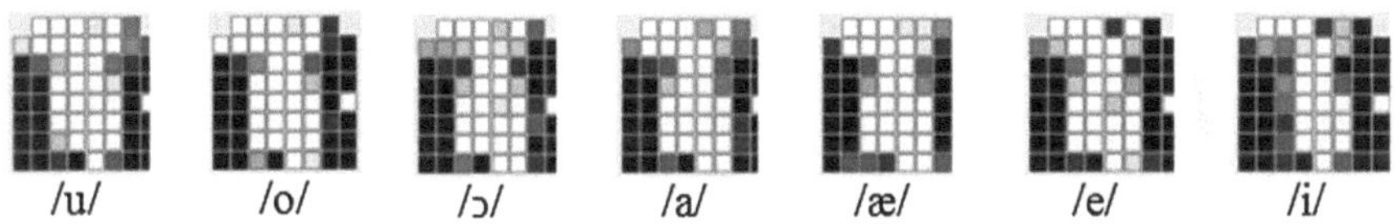

Fig. A.55 place of constriction of [ʤ] for seven Bangla vowels of female speaker

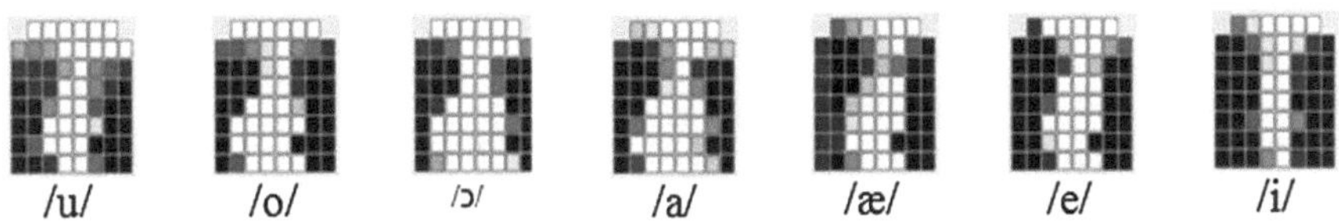

Fig. A.56 Place of constriction of [ʤ] for seven Bangla vowels of male speaker

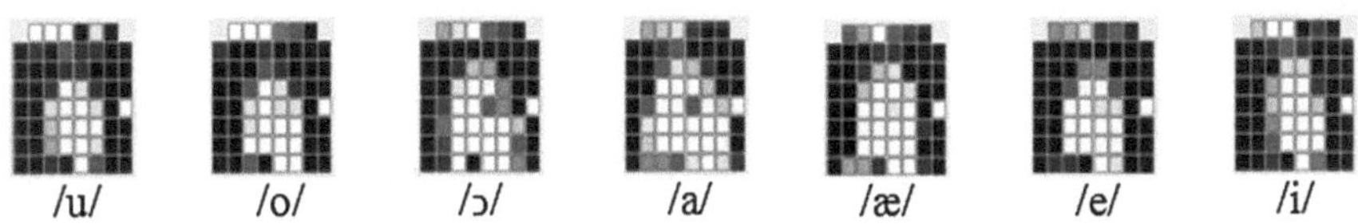

Fig. A.57 Place of closure of [ʤ^{h}] for seven Bangla vowels of female speaker

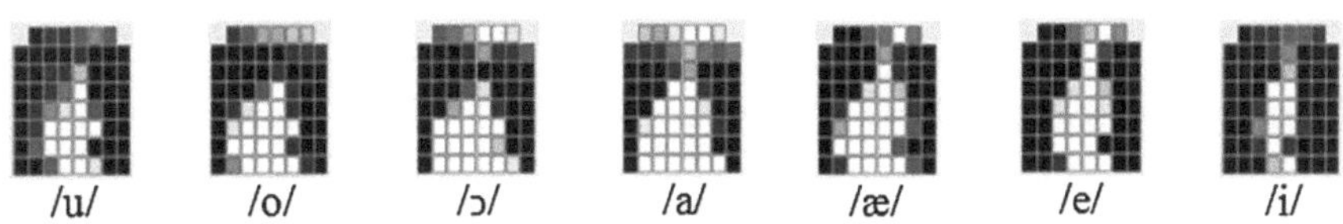

Fig. A.58 Place of closure of [ʤ^{h}] for seven Bangla vowels of male speaker

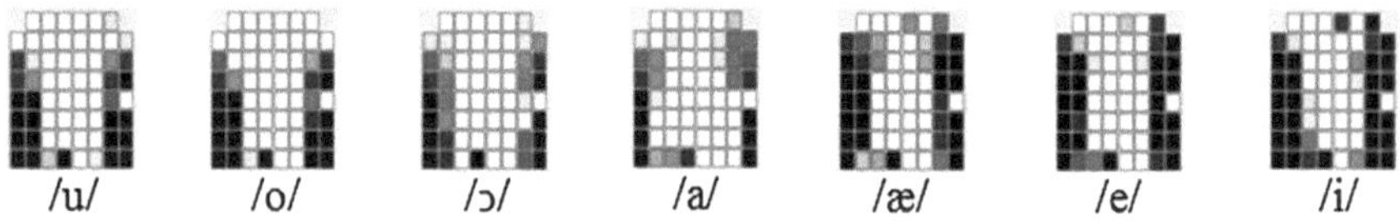

Fig. A.59 Place of constriction of $[ʤ^{h}]$ for seven Bangla vowels of female speaker

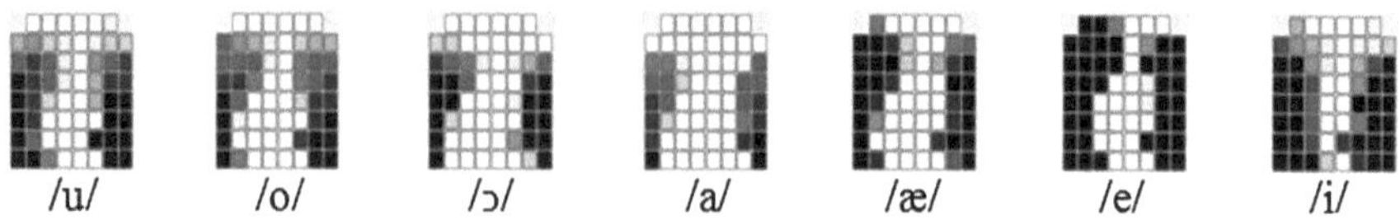

Fig. A.60 Place of constriction of $[ʤ^{h}]$ for seven Bangla vowels of male speaker

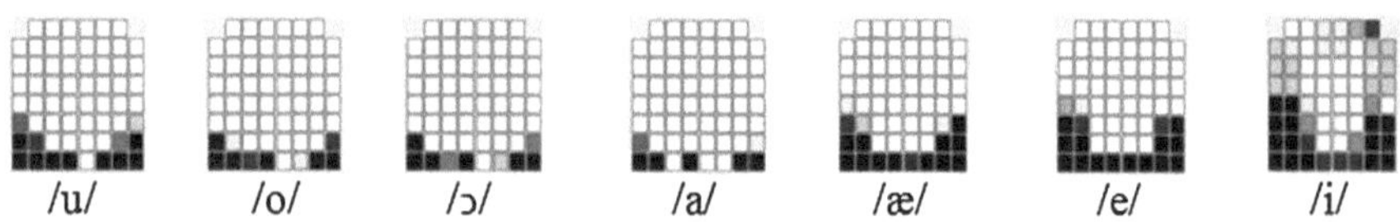

Fig. A.61 EPG frame of [ŋ] for all the seven vowel of female speaker

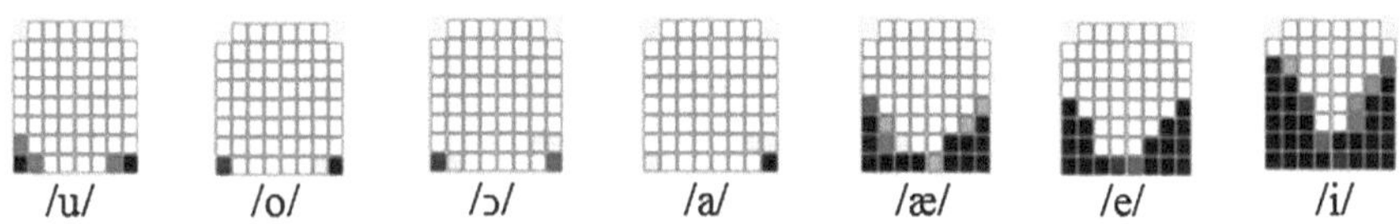

Fig. A.62 EPG frame of [ŋ] for all the seven vowel of male speaker

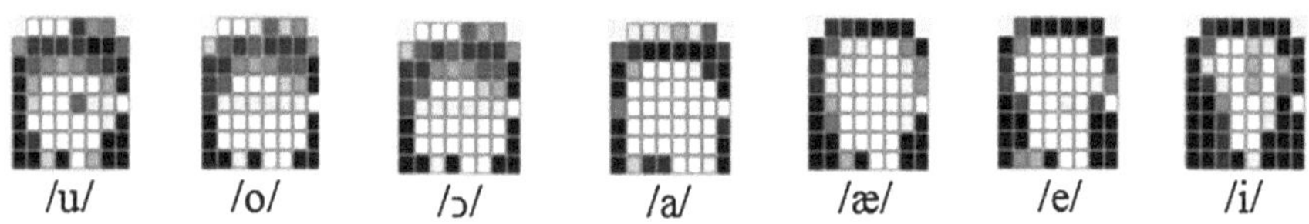

Fig. A.63 EPG frame of [ɲ] for all the seven vowel of female speaker

Constriction for male speaker is clearly visible, remarkably long for vowel /i/. Comparison of closure frame with those of constriction indicates retraction of tongue on release.

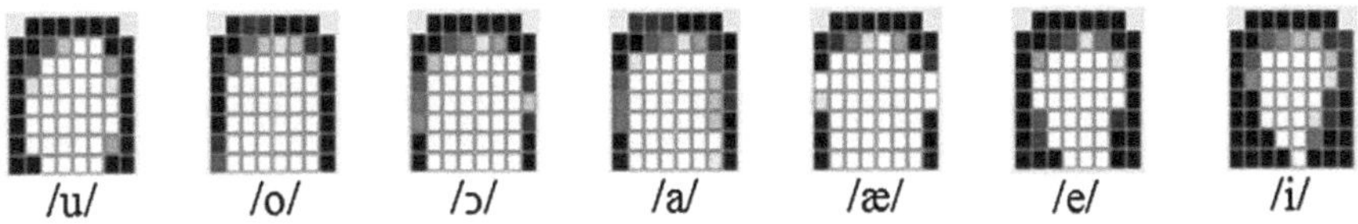

Fig. A.64 EPG frame of [ɲ] for all the seven vowel of male speaker

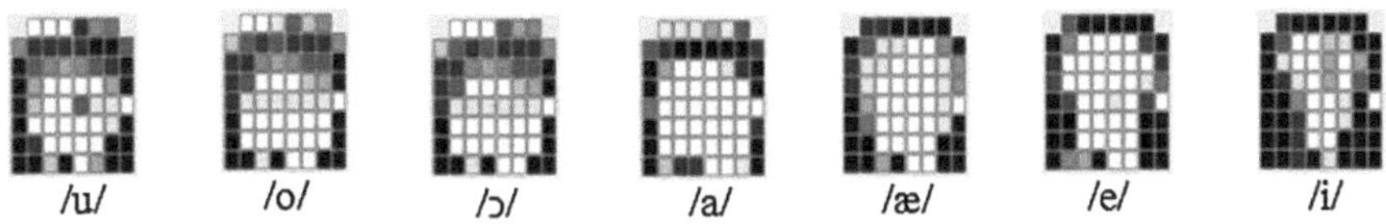

Fig. A.65 EPG frame of [n] for all the seven vowel of female speaker

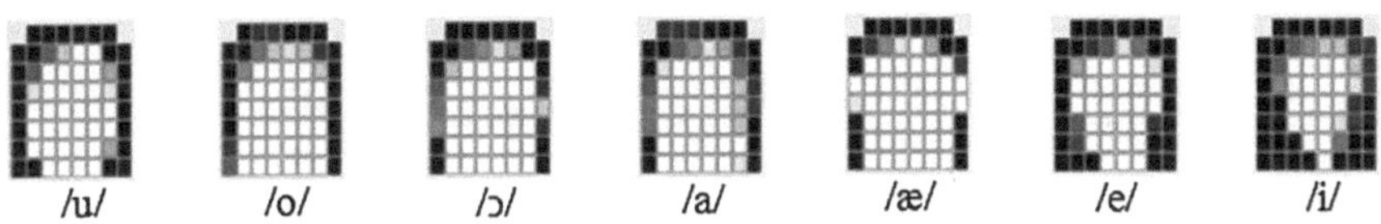

Fig. A.66 EPG frame of [n] for all the seven vowel of male speaker

Nasal consonants

See Figs. A.61, A.62, A.63, A.64, A.65 and A.66.

FSC
www.fsc.org
MIX
Papier | Fördert gute Waldnutzung
FSC® C083411

Kolibri 360 GmbH
Ferdinand-Jühlke-Straße 7
99095 Erfurt, Deutschland
produktsicherheit@kolibri360.de